the gadget geek's guide to

your Xbox 360™

By Jonathan S. Harbour
University of Advancing Technology

ISBN: 1-59863-173-x

Library of Congress Catalog Card Number: 2006920367

Printed in the United States of America

06 07 08 09 10 PH 10 9 8 7 6 5 4 3 2 1

THOMSON
™
COURSE TECHNOLOGY

Professional ■ Technical ■ Reference

Thomson Course Technology PTR, a division of Thomson Learning Inc.
25 Thomson Place
Boston, MA 02210
http://www.courseptr.com

Publisher and General Manager, Thomson Course Technology PTR:
Stacy L. Hiquet

Associate Director of Marketing:
Sarah O'Donnell

Manager of Editorial Services:
Heather Talbot

Marketing Manager:
Heather Hurley

Acquisitions Editor:
Mitzi Koontz

Marketing Coordinator:
Jordan Casey

Project Editor/Copy Editor:
Karen A. Gill

Technical Reviewers:
Bryan Hiquet, Parker Hiquet

PTR Editorial Services Coordinator:
Elizabeth Furbish

Interior Layout Tech:
DPS

Cover Designer:
Mike Tanamachi

Indexer:
Sharon Shock

Proofreader:
Carla Spoon

For my mother-in-law
Barbara Yoder
and my father-in-law
Dave Yoder

Acknowledgments

I thank God for the many opportunities that have come my way this year, such as the chance to write this book, and for the apparent talent needed to make something tangible of it.

I am grateful to my family for their ongoing encouragement: Jennifer, Jeremiah, Kayleigh, Kaitlyn, Mom and Dad, Grandma Cremeen, Dave and Barbara, and Grandma and Grandpa Schleiss; and for my extended family at Vision Baptist Church and Pastor Perham.

Thank you to the students, faculty, and staff at UAT for engendering such a wonderfully creative environment for learning.

Many thanks to the folks who made this book happen: Karen Gill, Mitzi Koontz, Bryan Hiquet, Parker Hiquet, Sharon Shock, Carla Spoon, and Mike Tanamachi.

About the Author

Jonathan S. Harbour is a senior instructor of game development at the University of Advancing Technology in Tempe, Arizona (http://www.uat.edu), where he teaches a variety of courses, from handhelds to consoles to game engines. He lives in the Arizona desert with his wife, Jennifer, and children, Jeremiah, Kayleigh, and Kaitlyn. You can reach him at http://www.jharbour.com.

Table of Contents

Table of Contents

Table of Contents

Table of Contents

Introduction

Are you a gadget geek or a hardware hacker? Do you feel compelled to learn everything there is to know about technology and how things work? Are you the expert that your friends go to when they have a problem with their computer, DVD player, MP3 player, surround system, or video game console? If so, you will enjoy this book, which explores the Xbox 360 inside out, with the terminology and expertise that you are used to. This book doesn't patronize you, and it was not written for a n00b. But if you are not a gadget geek, this book may at least get you going in the right direction.

There are 10 chapters in this book devoted to unlocking the full potential of your Xbox 360. The first six chapters show you how to make the most of your Xbox 360 console to do everything from playing games to viewing slideshows of your favorite digital photos to playing music streamed from your networked Windows PC. You will also learn about the accessories available for your Xbox 360, and I will show you how to connect your Xbox 360 to a home network so that you can play games online via Xbox Live. The remaining four chapters explore Xbox Live in detail, showing how to download game demos, movie trailers, themes, and pictures from Xbox Live Marketplace, how to download and play casual games from Xbox Live Arcade, and how to interact with other players using the Xbox 360's community features such as instant messaging, voice mail, and of course, online gaming.

- Chapter 1: Firing Up Your Xbox 360
- Chapter 2: Xbox 360 Media Player
- Chapter 3: Xbox 360 Games
- Chapter 4: Gadget Geek Gear: Xbox 360 Accessories
- Chapter 5: Connecting Your Xbox 360 to Your Home Network
- Chapter 6: Sharing Media Files over the Network
- Chapter 7: Going Online with Xbox Live

- Chapter 8: Xbox Live Marketplace: Game Demos, Trailers, Themes, and Microsoft Points
- Chapter 9: Xbox Live Arcade: Downloading "Casual" Games
- Chapter 10: Xbox Live Community: Ranking, Reputation, Voice Chat, Friend Lists, and Instant Messaging

Regardless of your experience level, you are bound to learn some interesting tricks that you may not have known were possible with an Xbox 360! (If you pay close attention, you may even learn how to drive the Nurbergring Nordschleife in under 20 minutes in a 2007 Shelby GT500– one of many tracks in *Project Gotham Racing 3!*)

STYLES USED IN THIS BOOK

The following styles are used in this book to highlight portions of text that are important. Beyond the Manual boxes provide more information about the subject without interrupting the flow of text. Geek Speak boxes provide definitions for technical terms. Meltdown boxes provide warnings for potential issues you may encounter with the technology. Here are some real examples from the book of these special sidebars.

Beyond the Manual

Chapter 4, "Gadget Geek Gear: Xbox 360 Accessories," goes over all the accessories that are available for the Xbox 360, including the optional hard drive.

Geek Speak

Ripping is a term that describes the process of copying music tracks from a physical disc to a digital file format such as MP3.

Meltdown

The Xbox 360 hard drive is required for playing original Xbox games. There is no workaround for this requirement. For this reason, the hard drive accessory is an inevitable upgrade if you don't own one already.

1

Firing Up Your
Xbox 360

The Xbox 360 (see Figure 1.1) is the most advanced video game console system ever developed, so it naturally caused quite a stir when it was launched on November 22, 2005. This date was momentous because it was the fourth anniversary of the launch of the original Xbox on November 22, 2001. The first Xbox launch was less eventful because it followed in the wake of the extraordinary success of the Sony PlayStation 2 video game system from the previous year. In addition, Nintendo followed with the release of the GameCube video game system just two weeks after the launch of Xbox.

To say that 2001 was an eventful year for the video game industry is quite an understatement. Sony was just starting to produce games that could take advantage of the PlayStation 2's (PS2's) hardware capabilities. Although Nintendo fans would have to wait a short time for a good selection of games for the new GameCube, Microsoft had pulled an ace from its sleeve with an extraordinary game to help launch the new Xbox. *Halo* not only helped Xbox sales, but it also became the standard for multiplayer console games and brought many new innovative ideas to market that had not been seen in a game in such a dynamic way before, including dynamic physics, intelligent friendly and enemy characters, a deep and immersive story, and outrageous, break-neck action that rivaled even the best PC first-person shooters at the time. Microsoft had not just entered into the console video game business—it had introduced a new console system with several launch games that were third-generation in quality.

Figure 1.1
The Xbox 360 is as powerful as it is attractive—quite a potent combination!

GENERATIONS

The video game business is a difficult one to be in, because the competition is fierce. But this business can also be profitable for a game producer and for the console manufacturer if a team is able to release the "Holy Grail" of the console market: a *killer game*. This term describes a game that is so innovative, compelling, or just plain fun to play that just about everyone who owns that console will buy the game in due time. *Halo* was the killer game that launched the Xbox into such great success in a crowded market (see Figure 1.2). You must keep in mind that Sega had been out-maneuvered by Sony just a year before the launch of the Xbox. The Sega Dreamcast was released in 1999, and it was a terrific video game console system with many innovative features and solid graphics performance. This console's lifetime was cut dramatically short by the release of the Sony PlayStation 2, which took away not just the newness that the Dreamcast had going for it, but also all of the game publishers.

Figure 1.2
Halo was a revolutionary game for a launch title.

Beyond the Manual

Are you a *Halo* uber-fan like me? If so, you should read the novels that have been written about *Halo*! If you want to know where the Master Chief came from before the events that took place in the game, read *The Fall of Reach*, by Eric Nylund. *The Flood* is the novelized version of the game, written by William C. Dietz. *First Strike* is the third novel in the series, and it delves into the events that happened after the end of *Halo*, leading up to the sequel, *Halo 2*.

Why did the Dreamcast fail so soon? Because the top game publishing companies want to max-imize the number of sales they will realize from a new game, and the best way to do that is to release games for the most popular system. The Sega Dreamcast might have beat Sony to market with a next-generation console by a whole year, but Sony had a far more powerful system in the PS2, and it offered backward compatibility with PlayStation games. Lagging sales led Sega to abandon production of the Dreamcast after only a two-year run. Today, Sega is a game devel-opment company rather than a hardware manufacturer. Table 1.1 shows the console system generations with the major players that helped define that generation. (I am aware of the many other consoles that played a role, but I am purposefully limiting the discussion for the sake of brevity.)

Table 1.1 Video Game Console Generations

Generation	Time Period	Console Systems
Precursors	Pre-1984	Magnavox Odyssey
		Atari VCS/2600
1st	1985–1991	Nintendo Entertainment System (NES)
		Sega Master System (SMS)
2nd	1991–1994	Super Nintendo (SNES)
		Sega Genesis/CD 32X
3rd	1993–1998	Atari Jaguar[1]
		Sega Saturn
		Sony PlayStation
		Nintendo 64
4th	1999–2005	Sega Dreamcast
		Sony PlayStation 2
		Microsoft Xbox
		Nintendo GameCube
5th	**2005–2010?**	Microsoft Xbox 360
		Sony PlayStation 3
		Nintendo Revolution

[1] The Jaguar came out during the transition into the 3rd generation, so it might be argued that the Jaguar belongs in the 2nd generation. However, the Jaguar's technical specifications place it in good company with 3rd-generation consoles.

A console video game system might be identified by a generational leap in processing power and graphics quality. The term *generation* is also used to describe the current level of quality available in the games. When a system is first launched, game developers have not had much time to learn how to maximize the capabilities of a console system, so the "launch title" games are typically not very impressive. Only after developers have had a couple of years to work out the problems, optimize the performance of their code, and learn the hardware performance tricks will the games start to reflect the true power of a system. When the PlayStation 2 was launched, the best thing it had going for it was the ability to play existing PlayStation 1 games. The PS2 was notoriously difficult to program, so it took more than a year of development time to come out with the "second generation" of games after the PS2 launch. The third-generation games typically reflect the true capabilities of a console system.

In retrospect, the Xbox featured several third-generation games at the time of launch that catapulted it far beyond what Sega had accomplished with its Dreamcast in a year. The reason that Xbox launch games like *Halo* were so advanced and fully developed was because the Xbox is similar to a PC. Microsoft gave Xbox developers an Xbox Development Kit (XDK) that looked and felt similar to the tools that PC game developers were already using: Microsoft Visual C++ compiler and the DirectX 8 game library. These tools were modified and customized just for the Xbox. This allowed Microsoft, for instance, to buy out Bungie and shift development of *Halo* to its new console system instead. That decision was a tipping point for the Xbox, which (arguably) might have gone the way of the Dreamcast without *Halo* and several other key games.

Hello, 360!

In light of the success and long life of the Xbox, for which games are still being developed at the time of this writing, Microsoft shifted gears for the Xbox 2. The first design decision was to come up with a better name than *Xbox 2*. Why was that necessary? I suspect this decision was a matter of marketing. Microsoft is all too aware of the pending release of the Sony PlayStation 3 a year hence. The last thing Microsoft wants is to wage a console system war with Sony, when that company's product *appears* to be more advanced, especially with the pending launch of Nintendo's next-generation system, code-named Revolution. When parents are browsing retail stores for a video game system, and they are not familiar with the different consoles, they are more likely to select the PS3 than an Xbox 2. So, a unique name was given to the Xbox 2, and that was Xbox 360. This new name accomplishes several goals from a marketing perspective. First, it includes the number 3 in the name. Second, it usurps Nintendo's upcoming system, Revolution. (You know, 360 degrees?) Third, the "360" moniker reflects Microsoft's goal to "surround" the gamer with an extraordinary gaming experience. (You will note the circles and arcs on all the accessory retail packaging.) This marketing tactic takes on both of the upcoming systems from Sony and Nintendo, thus "killing two birds with one stone," as the saying goes. Whether the Xbox 360 is more capable than its two upcoming competitors remains to be seen, but as you will learn in this chapter, the Xbox 360 is an extraordinarily powerful video game system that no doubt has Sony and Nintendo worried.

Microsoft has an interesting position to ponder with the new Xbox 360. Because Microsoft is the market leader for PC software and has been developing the popular Windows Media Center operating system—based on Windows XP—for several years now, the opportunity is ripe to build a new console system that can communicate with Windows operating systems running on PCs. That's exactly what Microsoft has done with the Xbox 360. As you shall learn in later chapters, the Xbox 360 can be used like a media hub for your living room or as a remote media player that shares resources (such as music, videos, and digital photos) with your Windows Media Center PC.

Of course, it's not about who makes a better console system. It's about which system has the best lineup of games. Nintendo has demonstrated for 20 years now that superior technology does not always deliver the most enjoyable games. The important factors are gameplay and design, not raw performance. The Nintendo Game Boy Advance is dramatically inferior to the Sony PlayStation Portable, yet it outsells the PSP consistently in hardware and games, and I haven't even brought the Nintendo DS into the equation. The Atari Jaguar was a revolutionary console system back in 1993. There were some very good games for the Jaguar, such as id Software's Doom (which was quite an accomplishment for a console at the time). Unfortunately for Atari, most game publishers were not interested in the hardware capabilities of the Jaguar because they were heavily involved in developing games for the Super Nintendo (SNES) and the Sega Genesis at the time. It was bad timing.

Timing Is Everything

In the video game console business, timing is everything. Then again, that rule applies to most industries. Follow your competition with every new product release, and you are bound to develop a reputation as a copycat and lose market share. On the other hand, trailing a competitor can work to your benefit if you design your system right and convince publishers to develop games for your system early on in its development.

Sega seems to have been caught in the unfortunate position of releasing an inferior console prematurely on two occasions. Sega released its Saturn system in 1994 for a whopping $399, taking advantage of the decline of the Atari Jaguar. One might say that the superior Saturn usurped the Jaguar in sales, but the truth is, SNES and Genesis systems were still selling well into 1994, so Sega had a strong source of income to help float its new Saturn system. The Saturn boasted a lot of new technology, including a CD-ROM system. Although Sega had released a 32-bit CD-ROM–based Genesis previously, the Saturn was the first next-generation system with a CD-ROM drive. Atari's Jaguar, on the other hand, still used the antiquated cartridge-based games, although it would eventually fall to the wayside due to weak game support.

Sega may have believed that the CD-based Saturn with a strong launch lineup of games (including the popular arcade game port of *Virtua Fighter*) would give Sega the market lead in the video game business. The real problem was that the Saturn was hobbled from the start. A fast pair of Hitachi SH2 32-bit RISC processors (operating at 28MHz) and a VDP1 32-bit graphics processor should have been up to the task, compared to the Jaguar and Sony PlayStation. Both the Saturn and PlayStation were equipped with 3MB of memory. The PlayStation featured an R3000A 32-bit RISC processor running at 33MHz, a 3D graphics chip, and a memory controller chip. The truth is, the Jaguar had comparable specifications. So why did the PlayStation outsell its two competitors by a wide margin? The answer may have a lot more to do with Sony's market penetration in other industries than with the quality and performance of the consoles.

The Sega Dreamcast was a fantastic console. It was well designed, innovative in every respect, and had a promising future. It was the first console system to include a modem (and later, an optional broadband adapter). Equipped with a 200MHz 128-bit Hitachi SH4 RISC processor, NEC CLX2 graphics processor, and 26MB of memory, the Dreamcast was a dream system for developers and gamers. Again, timing was a critical factor. If the Dreamcast had been released a year earlier (1998 instead of 1999 in the United States), it might have gained enough market share to compete with Sony's PlayStation 2 that was released in 2000. Sony was responsible for the short life of both the Saturn and the Dreamcast. The good news for gamers was that 250 games were published for Dreamcast in the United States during its three-year production cycle, which is far more games than most gamers own. So, in that respect, the Dreamcast was a resounding success that was simply cut short.

Three Modern Contenders

The console video game industry today is dominated by three products that vary in market share from one year to the next depending on the popularity of new games released. The PS2 owned about 60 percent of the market share, with Microsoft and Nintendo vying for second place in early 2003. By the second quarter of 2004, Microsoft took the market share from Sony and has retained that share for more than a year. Although the PS2 owned the video game market for four years, it was usurped by the newcomer to the industry, Microsoft, with Nintendo then coming in third behind Sony.

What speculation would I dare make now about the future of the console video game industry? Because Microsoft and Sony have swapped the lead position on market share of their fourth-generation consoles, the same may happen again during the fifth generation, of which Xbox 360 is the first member. I believe it's safe to say that the Xbox 360, Sony PlayStation 3, and Nintendo Revolution will be present and accounted for during the run of the fifth generation. However, competition will be fierce, and the level of game detail and the complexity of gameplay will far exceed everyone's expectations over the next few years.

XBOX 360 HARDWARE

The Xbox 360 has been designed as a high-definition game platform from the ground up. Although the Xbox supported HDTV resolutions of 480p, 720p, and 1080i, few games utilized high definition. The Xbox 360, on the other hand, requires this support from all games as a baseline. Those gamers who do not have an HDTV will still be able to play games on a standard

TV, but they will not get the same experience because many of the fantastic details made possible by HDTV will be filtered out for a standard TV due to the drop in resolution (down to just 320×240).

The Xbox 360's memory is top-notch, as is the rest of the hardware. The console has 512MB of onboard RAM. Now, this is not your typical 512MB double data rate (DDR) memory chip; the 360 uses GDDR3 RAM, with a memory clock speed of 700MHz. This high-caliber memory is normally reserved just for expensive graphics cards.

Central Processing Unit (CPU)

The central processor of the Xbox 360 was custom-designed by IBM just for this unique console, based on the PowerPC platform (the same processor family used in modern Apple iMac computers). This processor is far more powerful than anything being used in PCs at the time of this writing, featuring three cores that each operate at a clock frequency of 3.2GHz. Most PCs and console systems today have a processor with just a single core. The *core* describes the part of the CPU that does all the processing for the system. Some computers are equipped with two processors, and they are called dual-processor systems. It is the job of the operating system to utilize the two processors to best run the software.

To understand how this works, you must become familiar with a software trick called multi-threading. All computer programs have a single thread, or process, that does whatever the program is supposed to do. But on modern operating systems (such as Windows and Mac OS X), which can run many programs at the same time, all of the tasks are divided into different threads. These threads are then run on each of the processors in a dual system. Some server computers have 4 processors, and a few have as many as 8 or 16 processors!

Each core of this high-performance PowerPC chip includes a VMX-128 vector unit. Each of these three vector units provides one of the six hardware threads with 128 floating-point registers, each of which is capable of manipulating (moving, rotating, or scaling) five polygons at a time. That's 640 polygons per thread, or 3840 total polygons every clock cycle. Note that I'm not referring to polygons *per second*, but *per clock*. The processor clock frequency is 3.2GHz, or billion hertz. Do the math, and you'll find that this processor is capable of calculating in the teraflop range. (That's more than a trillion operations per second!) How is Microsoft able to sell the Xbox 360 for $299 with this kind of hardware inside? I would love to have this processor in my PC, and I don't even have dual processors, let alone a Pentium D or a dual-core Athlon 64.

Intel and AMD have recently developed dual-processor chips that have the equivalent of two processors on a single chip. This would have required a special motherboard with two processor sockets in the past, but now you can purchase a new chip that includes both processors on the same chip. This is a fairly recent microprocessor innovation, but it gets better: Each core in the Xbox 360 processor can handle two threads at the same time, for a total of *six* hardware threads simultaneously! What does this mean in real-world figures? This processor can crunch numbers at a rate of 115 gigaflops (billion floating point operations per second). That IBM was able to mass-produce a custom PowerPC with a *triple* core (each running at 3.2GHz) is nothing short of extraordinary. This processor might well be the *coup de grace* that solidifies Microsoft's position as market leader of the next-generation console video game industry.

Graphics Processing Unit (GPU)

The Graphics Processing Unit (GPU) used by the Xbox 360 is a custom-made chip manufactured by ATI Technologies, which is the number-one graphics chip manufacturer in the world. (nVidia is close on its heels.) This custom ATI chip is comparable to a Radeon X800 with 10MB of texture memory built into the chip, operating at a core clock frequency of 500MHz. The GPU has a 48-way pipeline operating in parallel to execute floating-point vertex and pixel shaders based on the unified shader architecture. This chip works to display polygons that have been manipulated by one of the vector units in the CPU for outstanding graphics realism never before seen in a game. The true capabilities of this hardware will not be realized in the launch title games, but rather only after developers have had time to get used to this extraordinary power.

That said, there are several Xbox 360 launch titles that do take advantage of this processing power. *Project Gotham Racing 3* features car models with an average of 90,000 polygons each. Most *entire games* don't feature nearly that many polygons!

POWERING UP YOUR XBOX 360 FOR THE FIRST TIME

The Xbox 360 is simple to set up if you don't mind using it with all of the factory default settings. As a gadget geek, that notion should offend you! So let's explore the 360 from initial startup to creating a player account to playing some games.

When you first power up your Xbox 360, you are presented with the following screen (see Figure 1.3). This screen is simple enough. You will probably just select the default option of English and move along.

Beyond the Manual

Do you have the Xbox 360 Media Remote or Universal Remote? If so, you might find it more convenient than a controller, especially when you are navigating the user interface. I find that the arrow buttons and large OK button are more convenient on the Media Remote than using the left and right hat and D-pad buttons on the controller.

Figure 1.3
The first screen you see when you power up the Xbox 360 lets you choose a language.

Next, you will see the Gamer Profile screen (see Figure 1.4). This screen, like many others in the Xbox 360's user interface, is similar to the installation wizards you often find within the Windows operating system. There are three options on this screen:

- **Create an offline profile.** I recommend that you start with this option, even if you already have an Xbox Live account (which is likely if you own an original Xbox). This option lets you explore the 360's user interface to see what the Xbox 360 has to offer by way of features. You will be able to sign onto Xbox Live at any time in the future.

- **I am a member of Xbox Live.** This option is for current Xbox Live subscribers. Selecting this option allows you to log into Xbox Live to retrieve your gamer profile, which is then downloaded to your 360.

- **I want to join Xbox Live.** This is the option to choose if you are eager to start playing games online right away. (As a true gadget geek, aren't you compelled to see what this thing can do before playing online?)

If you choose the first option to create an offline profile, you are presented with the Gamer Profile Name screen, allowing you to type in your nickname (see Figure 1.5). Use the controller D-pad or remote control arrows to move the cursor, and press A or OK on the remote to select letters. The X button performs a backspace.

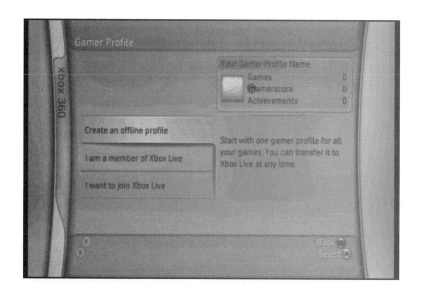

Figure 1.4
The Gamer Profile screen.

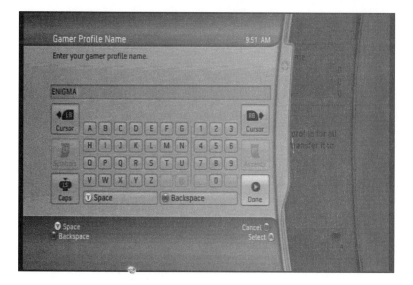

Figure 1.5
The Game Profile Name screen.

The Gamer Picture screen comes next (see Figure 1.6). You must select an avatar to represent your icon image. You use this avatar everywhere you go, and it's especially important when you start using the instant messaging (IM) feature of Xbox Live to chat with other players online, so choose a good image that reflects your style.

Figure 1.6
The Gamer Picture screen.

Beyond the Manual

Just how much of a gadget geek are you? If you haven't yet tried to plug a USB keyboard into your Xbox 360 to simplify these difficult hunt-and-peck data entry screens, I'm not sure there's hope for you! Yes, the Xbox 360 supports USB keyboards. Believe it or not, the console even works with non-Microsoft keyboards (although I won't confirm or deny that). I recommend a small wireless keyboard with a USB receiver for the ultimate gadget *connoisseur*. If you want to impress your friends, use a white wireless keyboard with charcoal-gray keys to match your 360 to the letter (pun intended).

Family Settings

One more screen is left, after which your Xbox 360 will be configured to your specifications. The Initial Setup Complete screen (see Figure 1.7) shows the settings you chose and gives you additional options that you often need during initial setup. These options are as follows:

Family Settings. You can use this option to limit access to games, movies, and online games based on ratings. This is especially helpful if you have younger gamers in the household who may not be old enough to play some types of games or watch certain movies (either played directly on the DVD-ROM drive or stored digitally on your Media Center PC or on the Xbox hard drive itself). Selecting this option brings up the Family Settings dialog box (see Figure 1.8). Two options are available here: Console Controls and Xbox Live Controls.

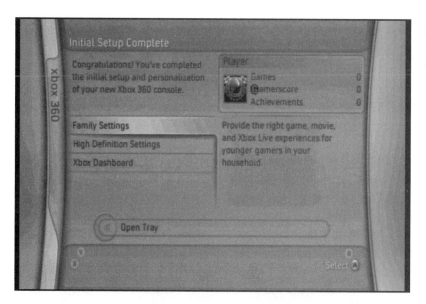

Figure 1.7
The Initial Setup Complete screen.

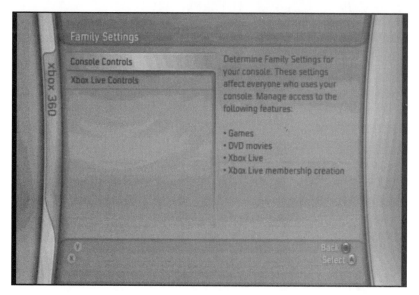

Figure 1.8
The Family Settings screen allows you to restrict access to games and movies based on content ratings.

There are seven options in the Console Controls screen (see Figure 1.9):

🔧 **Game Ratings.** This option allows you to restrict access to games based on the Entertainment Software Rating Board (ESRB) content ratings (see Figure 1.10). The ESRB defines the following video game ratings:

EC—Early Childhood

E—Everyone (6 and older)

E10+— Everyone (10 and older)

T—Teen (13 and older)

M—Mature (17 and older)

Figure 1.9
The Console Controls screen.

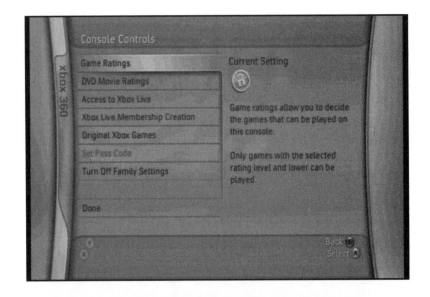

Figure 1.10
The Game Ratings screen.

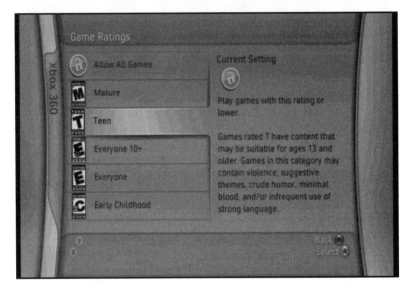

- **DVD Movie Ratings.** This option allows you to restrict access to DVD movies based on Classification and Rating Administration (CARA) content ratings (see Figure 1.11). CARA defines the following movie ratings:

  G—General Audiences

  PG—Parental Guidance Suggested

  PG-13—Parents Strongly Cautioned

  R—Restricted

  NC-17—Not Censored/Adults only

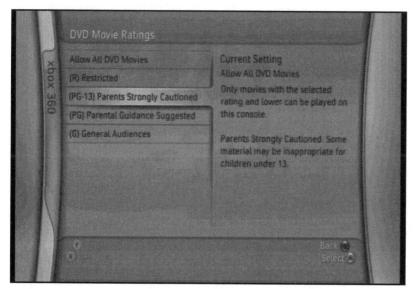

Figure 1.11
The DVD Movie Ratings screen.

- **Access to Xbox Live.** This option lets you specify whether players can go online.

- **Xbox Live Membership Creation.** This option allows you to restrict the ability to create new Xbox Live accounts.

- **Original Xbox Games.** This option determines whether you can play old Xbox games on your 360.

- **Set Pass Code.** This option sets a "parent password" to save the changes made to the Family Settings.

- **Turn Off Family Settings.** This option disables all Family Settings.

High-Definition Settings

Back at the Initial Setup Complete screen, the second option allows you to configure your Xbox 360's high-definition settings (see Figure 1.12). If you don't own an HDTV, you can ignore this screen, but I will go over it briefly now anyway for those who are lucky enough to have an HDTV. Although the Xbox 360 was designed from the ground up with high definition in mind, it still supports older TV technology.

Figure 1.12
The Console Settings screen.

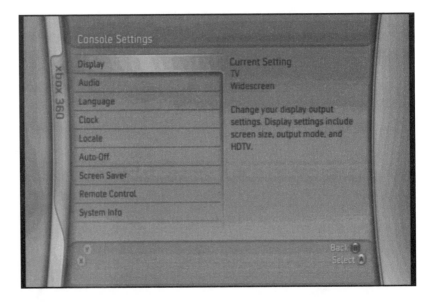

You can change the high-definition settings by using the Console Settings screen from the main Xbox Dashboard, so this screen is shared during initial setup. There are nine options in this screen, and I recommend going through them to configure your Xbox 360 completely. For instance, this is where you set the current date and time. The clock resets if you unplug your Xbox 360, but the other settings (such as time zone and display settings) are retained in the Xbox 360 BIOS chip, which is a solid-state EEPROM (Electrically Erasable Programmable Read Only Memory) chip. What this means is that all of your configuration settings are saved on the Xbox 360. The only exception is the real-time clock, which obviously can't continue to keep time when there is no power.

Here are the nine Console Settings screen options:

- **Display.** The Display Settings screen allows you to change the HDTV modes (480p, 720p, or 1080i) and screen format (normal or widescreen).

- **Audio.** The Audio Settings screen is where you select analog output (Dolby Pro Logic II or mono) or digital output (Digital Stereo or Dolby Digital 5.1 with a WMA Pro option for pass-through audio).

- **Language.** The language screen is used to change the language in the Dashboard.

- **Clock.** The Clock Settings screen allows you to set the date and time.

- **Locale.** This is the geographic location in which you live.

- **Auto-Off.** This controls whether the 360 will shut itself down after a certain number of hours of inactivity.

- **Screen Saver.** This controls whether the screen saver should dim the screen.

- **Remote Control.** This configures the type of remote control you use.

- **System Info.** This option displays the console serial number, console identifier, and BIOS revision number. Write down the serial number and console identifier for warranty purposes (should the need arise). The BIOS version is the number at the bottom of this screen. At the time of this writing, the latest Xbox 360 BIOS revision is 2.0.1888.0.

Xbox Dashboard

The third and final option in the Initial Setup Complete screen takes you to the Xbox Dashboard. The Dashboard is a new concept for a video game console that was introduced with the original Xbox, representing the "desktop" on a Windows PC. The Xbox Dashboard is similar to the user interface of the Media Center Edition of Windows, providing a simple navigation system that you can control using a remote control (as in the case of Media Center PC) or a standard video game controller. The Dashboard is shown in Figure 1.13. We will explore the Dashboard extensively in the next three chapters.

Figure 1.13
The Xbox 360
Dashboard.

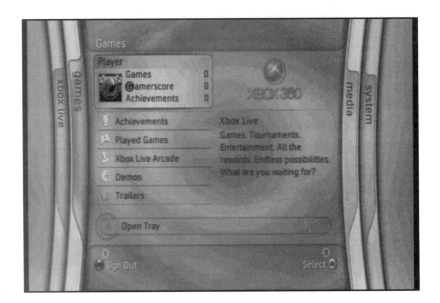

the
gadget
geek's
guide to
Your Xbox 360

2

Xbox 360 Media Player

This chapter explores the most significant capability of the Xbox 360. Although your first inclination might be to suggest that the Xbox 360 is all about games, I believe this chapter will prove to you that the Xbox 360 was designed primarily as a media player and only secondarily as a video game console. I will show you how to connect directly to some of your media devices using your Xbox 360 to play a variety of media files, even if you don't own the Xbox 360 hard drive accessory.

RICH MULTIMEDIA EXPERIENCE

The Xbox 360 brings a rich multimedia experience to your living room. Although the term *multimedia* has fallen out of style in recent years, it nevertheless describes the capabilities of the Xbox 360. You can play music files stored in popular formats, CD audio, and DVD audio—even while doing other things such as viewing photos, chatting with friends online, or playing games. This is possible due to the powerful processing power of the Xbox 360 (which we gawked over in the previous chapter).

Thinking Too Hard?

Of course, even a casual PC user is already familiar with the term *multitasking*. This term describes the situation when you do more than one thing at a time. Most of us are not able to count to 100 in our heads while also reading a book or studying for class, because these are difficult thought processes that require our attention. You can probably do a few minor mental tasks simultaneously, but not for an extended period of time.

However, it's a completely different story when it comes to sensory input. Our minds are capable of sifting through vast amounts of input to identify patterns, which is what the human mind is best at doing—pattern recognition. It requires almost no effort to listen to your favorite music (and sing along) while playing a game. The Xbox 360 was designed to take advantage of these aspects of human-machine interaction so that your experience is seamless. You will need no more thought to navigate the Xbox 360 interface than you spend singing along with a favorite song.

Xbox 360 Dashboard

Figure 2.1 shows the Xbox 360 Dashboard, which is the primary interface to the operating system. It's comparable to Windows Explorer and the Start menu in a Windows XP system. If you use Windows XP Media Center Edition, the interface should be familiar to you already. There are four folders in the Xbox Dashboard:

- System
- Media
- Games
- Xbox Live

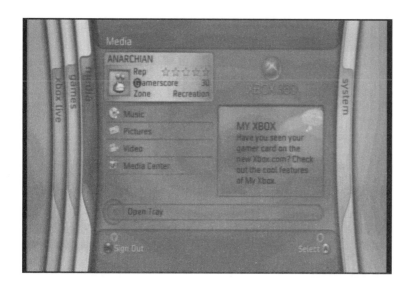

Figure 2.1
The Xbox 360 Dashboard with the Media folder highlighted.

At the top of every page in the Dashboard is a small window showing your gamer profile information (which is currently highlighted in Figure 2.1). If you select your profile (by pressing OK on the Media Remote or A on the controller), the Gamer Profile window opens, shown in Figure 2.2. This window displays all the details about your gamer profile, including your reputation, your Gamerscore (a value indicating the number of accomplishments achieved during gameplay, both online and offline in solo games), and your preferred zone.

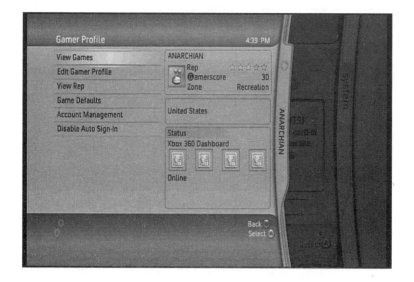

Figure 2.2
The Gamer Profile window.

Beyond the Manual

I spend more time talking about gamer profiles and the other features of online play in Chapter 7, "Going Online with Xbox Live." Until then, I will keep the discussion centered on offline gameplay because you may not have signed up to play online yet.

For now, back out of the Gamer Profile window by pressing B on your controller or Back on your remote to return to the Media folder interface. As you can see, there are four options on this screen:

- Music
- Pictures
- Video
- Media Center

The fourth option in the Media folder of the Dashboard, Media Center, is used when you want to connect to a Windows PC on your network to share media files with the 360. This somewhat advanced subject—which requires configuration of your Windows PC to work with the 360— is reserved for Chapter 5, "Connecting Your Xbox 360 to Your Home Network."

PLAYING MUSIC WITH YOUR XBOX 360

Move the selector down to highlight Music, as shown in Figure 2.3.

Figure 2.3

The Music option allows you to listen to music on your Xbox 360.

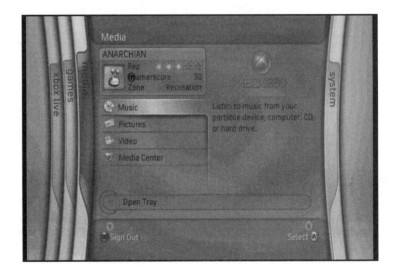

Selecting the Music option opens the Music Player window shown in Figure 2.4. When you enter this screen for the first time, the Music Player option is not available because you have to select a music source first. The sources are shown in the list and include the following:

- Hard Drive
- Computer
- Current Disc
- Portable Device

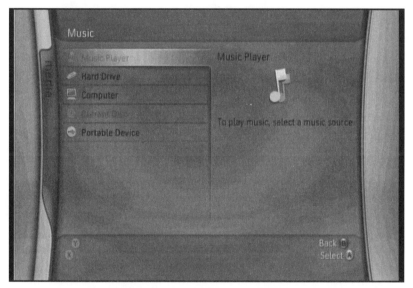

Figure 2.4
The Music Player can play music files from a variety of sources.

Playing Music from a Portable Device

You can select music that may already be available on the hard drive if your Xbox 360 is so equipped. Normally, the Portable Device option is dimmed, but in this case I have a USB flash stick inserted into one of the Xbox 360's USB ports on the front of the case (see Figure 2.5).

Beyond the Manual

Chapter 4, "Gadget Geek Gear: Xbox 360 Accessories," goes over all the accessories that are available for the Xbox 360, including the optional hard drive.

Figure 2.5
You can plug USB flash sticks and other memory card devices into the Xbox 360.

The Xbox 360 supports just about every USB flash stick on the market, from the early 32MB sticks to the latest 1GB models (see Figure 2.6). And the 360 is not limited to just USB sticks; it also supports most USB memory card adapters, from a single adapter to a multicard adapter supporting CompactFlash, SmartMedia, Secure Digital, MMC, and other formats. As long as you

Figure 2.6
An assortment of common USB flash drives that are compatible with the Xbox 360.

have a USB interface and the device supports Plug-N-Play (meaning it can be used on your Windows PC without any driver software), the 360 should be able to use it.

If you have a USB flash stick or some other USB memory card adapter available, copy some of your music collection to that device using your PC so that you can try playing music from that portable device. When you insert a USB device into one of the USB ports, the Media Player window refreshes and shows that the Portable Device option is now available. Move the selector down to Portable Device and select it, as shown in Figure 2.7.

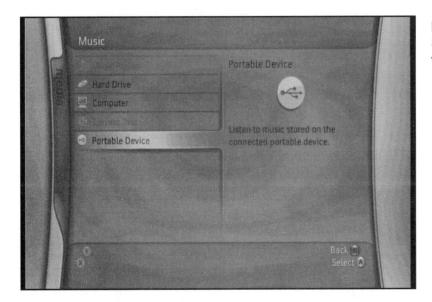

Figure 2.7
Selecting a portable device for music playback.

When you select this option, the Portable Device window opens. You see the files that Media Player detected on the portable device that it can play. As shown in Figure 2.8, I have two albums on this USB flash stick.

By moving the pointer to an artist in the list (on the right side), I can open the albums that are available for that artist. (Media Player organizes the list for you.) By selecting Lenny Kravitz, as shown in Figure 2.9, you can see that album 5 is displayed.

Figure 2.8
Media Player organizes
music files by artist and
then by album.

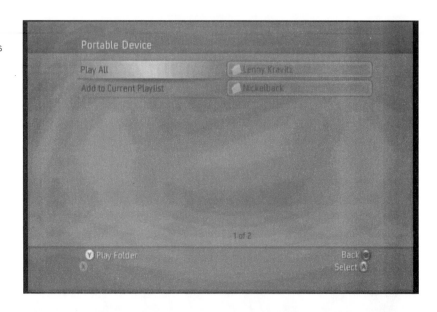

Figure 2.9
The albums are listed for
the currently selected
artist.

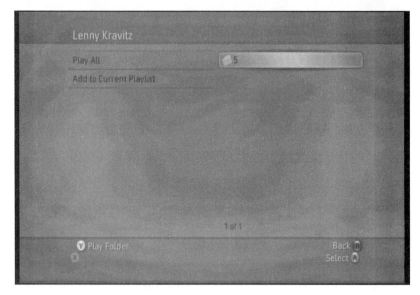

Now I can either select Play All or add the album to my current playlist. (Yes, Media Player
supports multiple music playlists.) I can also select the album name to view the song tracks in
the album. By opening the album, I can see the songs, and I have the option to add individual
songs to my playlist (see Figure 2.10).

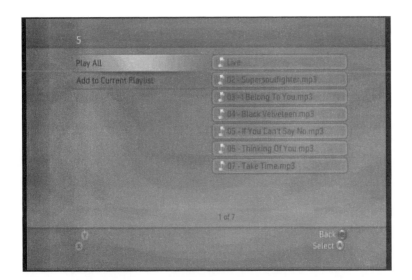

Figure 2.10
Viewing the song tracks for an individual album.

Media Player reads the information in each music file to determine the artist name, album name, and other details of the song. For instance, when you select a song in the album view window, the song detail window comes up, showing details about the song (see Figure 2.11).

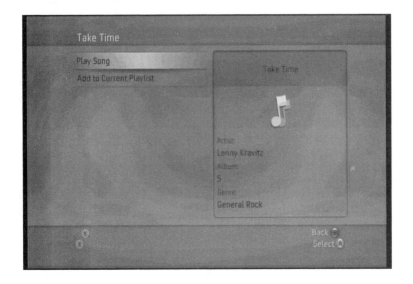

Figure 2.11
Viewing the details of a single song.

When you select the Add to Current Playlist option, the song is added to your current playlist. As you can see in Figure 2.12, I have only one song in my playlist so far. I can continue browsing files on my portable device, on a music CD, or on the Xbox 360 hard drive and add additional songs to my playlist. Just keep in mind that removing a portable device causes links to songs in your playlist to be removed, too. Note the Edit or Save Playlist option, which you can use to manage your playlists.

Figure 2.12
The Music Player screen shows your current playlist.

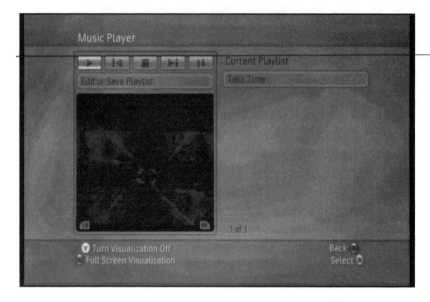

Playing and Ripping Music from a CD

Let's go back to the main Music menu by pressing the B button. Note that the Music Player option is no longer dimmed because I have just added some songs to my playlist that may still be playing "in the background." (Remember that the Xbox 360 is a multitasking console.) I can open the Music Player if I want to manage my playlist, skip tracks, or stop the music.

Beyond the Manual

You don't need to open the Music Player to control music playback if you have a Media Remote. Both the Premium and Universal remotes have buttons for controlling music playback.

Now let me show you how easy it is to play a music CD. Insert a music CD into the DVD tray of your Xbox 360. This causes the Music Player to open automatically and start playing your music CD (see Figure 2.13).

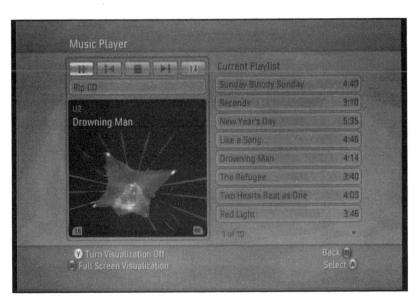

Figure 2.13
The Music Player comes up automatically when you insert a music CD.

If I return to the music options screen, I notice that Current Disc is no longer dimmed and is available for use (see Figure 2.14). By using this interface, I can browse the songs on a CD and play them directly from there, or I can rip them to the hard drive.

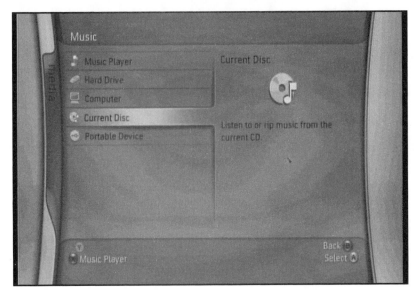

Figure 2.14
The Current Disc option is now available in the Music screen.

Geek Speak

Ripping is a term that describes the process of copying music tracks from a physical disc to a digital file format such as MP3.

Beyond the Manual

Your Xbox 360 downloads the album information for a CD you insert into the drive only if you have access to Xbox Live. This is an essential feature for ripping music because it is impossible to organize your music without album information. If you don't have an online account, you have to edit the album information manually (which is where a USB keyboard is a time-saving commodity).

Take a look at the Rip CD option located just above the visualizer window back in Figure 2.13. You must use this feature (you need to have a hard drive, too) to assemble a playlist with the songs on a music CD, because you can't add CD tracks to a playlist—you can only play a music CD directly. Selecting the Rip CD option brings up the screen shown in Figure 2.15.

Figure 2.15
Your Xbox 360 can rip music CDs for use in your music playlists.

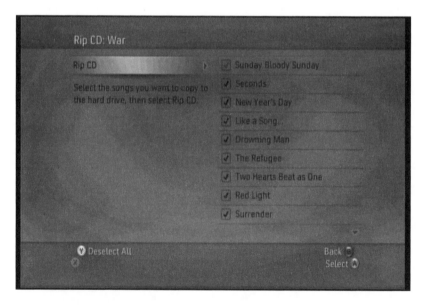

You can select or deselect individual songs from the music CD to rip to the hard drive. Select the Rip CD option when you're ready to go. The process can take a while because the CD tracks must be read and encoded into a compressed music file format. The ripping process is shown in Figure 2.16.

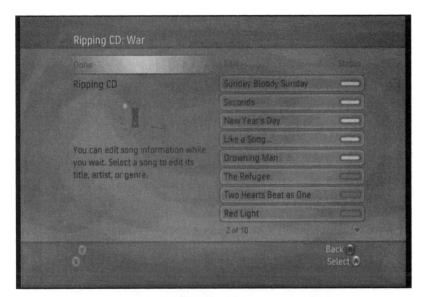

Figure 2.16
Ripping a music CD to the hard drive.

Beyond the Manual

You cannot rip a CD if your Xbox 360 is not equipped with a hard drive, because memory units can only be used for storing saved games. Also, you cannot rip a CD to a portable device. (Isn't that what your PC is for?)

When the CD has been copied to the hard drive, you see the screen shown in Figure 2.17. The music CD tracks are now available for use in playlists.

Accessing Music on the Hard Drive

The hard drive (see Figure 2.18) is so integral to the Xbox 360 experience that it's hard to imagine getting along without one. When you consider that a typical memory unit (see Figure 2.19) costs around $40, the price of the hard drive seems reasonable. After all, a memory unit holds only 64MB, whereas the hard drive holds 20GB. That's 312 times more storage space for the price of four memory units!

Figure 2.17
The CD has been successfully copied to the hard drive.

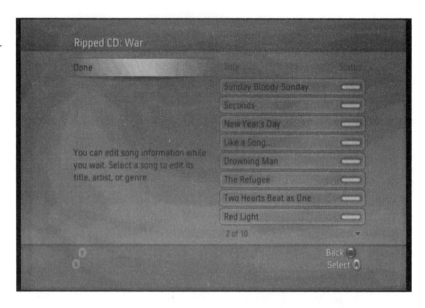

Figure 2.18
The Xbox 360 hard drive holds 20GB of data.

Figure 2.19
The Xbox 360 memory
unit holds 64MB of data.

If you have a hard drive installed in your Xbox 360, go to the main Music screen and select Hard Drive, as shown in Figure 2.20.

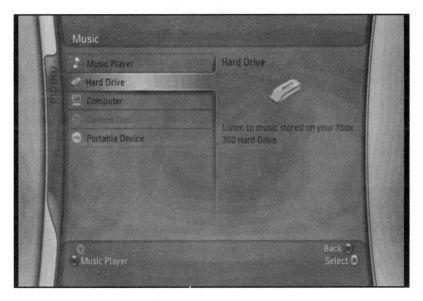

Figure 2.20
The Hard Drive option is
available in the Music
menu if your 360 is
equipped with a hard
drive.

The Hard Drive screen (see Figure 2.21) is basically a music browser that allows you to sort through your music collection using the following search criteria:

- Albums
- Artists
- Saved Playlists
- Songs
- Genres

Figure 2.21
Browsing your music col-
lection that's stored on the
hard drive.

The hard drive is preloaded with numerous sample songs from a variety of bands that you can use in a custom playlist while you're playing games. Selecting an album brings up the screen shown in Figure 2.22, which gives you the following options:

- Play Album
- Add to Current Playlist
- Edit Album Info
- Delete Album

Streaming Music from a Windows PC

One of the most impressive features of the Xbox 360 is its ability to access media on a Windows PC over a Local Area Network (LAN). If you have Windows XP Media Center Edition on a PC that is connected to the same network as your Xbox 360, you can stream TV to your Xbox 360 and access the full suite of features available on your Media Center PC, including photos, music, and videos.

Even if you don't have a Media Center PC, you can still use a standard Windows XP system to share media files with your Xbox 360. However, first you must install a special server program on the PC that allows the Xbox 360 to access media files on it. I fully cover this subject in Chapter 5, so I'll hold off on the details here for now.

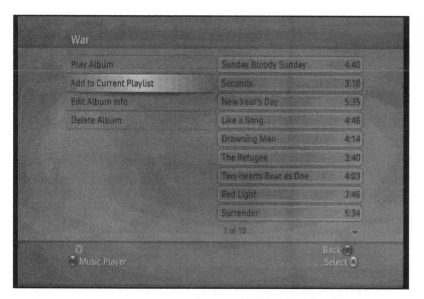

Figure 2.22
Browsing your music collection that's stored on the hard drive.

Beyond the Manual

The media-sharing capabilities of the Xbox 360 are a one-way relationship; your PC cannot access the media files on your Xbox 360 because the system is secured from tampering for security reasons. If we had access to the 360's file system, we could hack it, which is why your 360 has no shared network resources.

Using the Media Remote

Up to this point, I have been using the Xbox 360 Media Remote almost exclusively for navigating the Dashboard interface. The Remote is far more convenient than using a controller when you want to do just about anything except play games, which is where the controller obviously shines.

Figure 2.23 shows the Universal Media Remote with a description of the groups of buttons. Another smaller Media Remote was provided with the Xbox 360 Premium retail packaging, lacking the TV tuner control and channel buttons.

Figure 2.23
The Xbox 360 Universal
Media Remote.

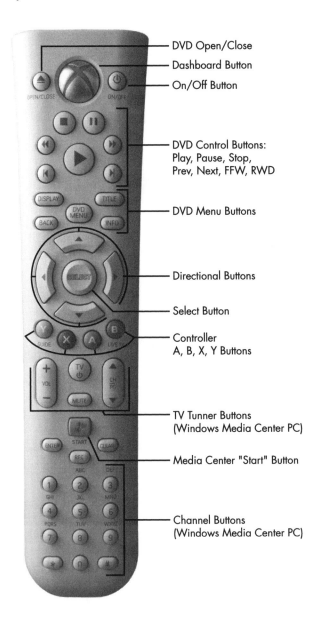

DVD Open/Close

Dashboard Button

On/Off Button

DVD Control Buttons:
Play, Pause, Stop,
Prev, Next, FFW, RWD

DVD Menu Buttons

Directional Buttons

Select Button

Controller
A, B, X, Y Buttons

TV Tunner Buttons
(Windows Media Center PC)

Media Center "Start" Button

Channel Buttons
(Windows Media Center PC)

VIEWING DIGITAL PHOTOS WITH YOUR XBOX 360

Your Xbox 360 can handle digital photos from a variety of sources:

- Computer
- Digital camera

📀 Current disc

📀 Portable device

To work with digital photos, open the Media folder in the Dashboard and select the Pictures option, as shown in Figure 2.24.

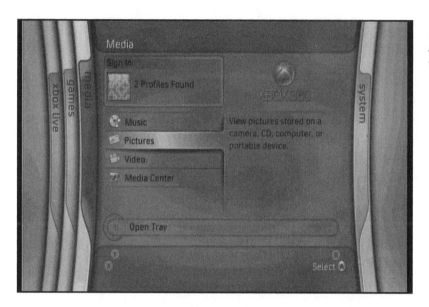

Figure 2.24
Accessing digital photos from the Media folder.

Selecting the Pictures option brings up the Pictures screen (see Figure 2.25). As you can see in this example, most of the options are dimmed out and unavailable; this is because I have not inserted devices into the 360 yet.

Viewing Photos on a Portable Device

Before you can view photos on a portable device such as a USB stick or media card reader, you need to copy your photos to the memory device using your PC. If you have never before used a memory card or thumb drive, I encourage you to buy one because they are invaluable. The current models give you a gigabyte (1GB) of storage for about $60.

Beyond the Manual

You will learn how to access digital photos over the network from your Windows XP or Media Center PC in Chapter 5.

Figure 2.25
The Pictures screen shows the sources available for viewing digital photos.

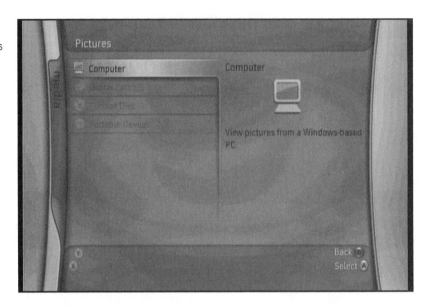

When you insert a portable device into one of the USB slots on the 360, you see the Portable Device option become available in the Pictures screen, as shown in Figure 2.26.

Figure 2.26
Portable Device is now available as an option in the Pictures screen.

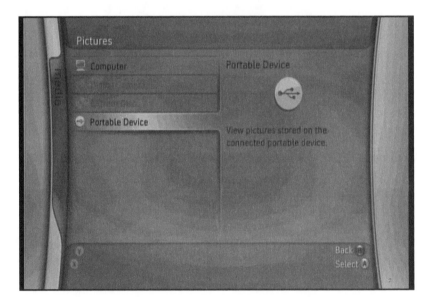

Selecting Portable Device opens the memory card device you have inserted and allows you to browse the photos on the device. Figure 2.27 shows the contents of a USB flash stick that I have inserted into one of the front USB ports on the 360.

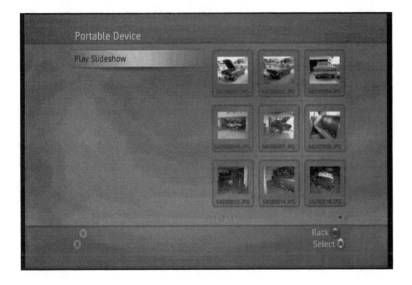

Figure 2.27
The contents of a USB flash stick are displayed in the Portable Device screen.

Using the directional buttons, I can navigate among the photo files and select one to view it full screen, as shown in Figure 2.28. When you have a photo on display, you can press one of the directional buttons to bring up a control menu with the following options (see Figure 2.29):

- Pause (slideshow)
- Go to Previous Image
- Stop (slideshow)
- Go to Next Image
- Toggle Shuffle On/Off
- Toggle Repeat On/Off
- Rotate Image Clockwise
- Rotate Image (counter-clockwise)

Figure 2.28
Viewing a digital photo
that's stored on a memory
stick.

Figure 2.29
The Photo Slideshow
control panel menu.

There are two options for rotating an image clockwise or counter-clockwise, which may be handy
if you happen to have digital photos that were taken in portrait mode (by holding the camera
sideways). Figure 2.30 shows the results.

Figure 2.30
Rotating a digital photo.

Beyond the Manual

When browsing the thumbnails of digital photos, you can press X to set a particular photo as the background image for your Dashboard.

From the thumbnail viewer screen, you can select Play Slideshow to view all the photos in a slideshow. This is a really nice feature, especially if your 360 is hooked up to a large-screen HDTV. The photos do look nice on any normal TV screen, though.

Meltdown

Do you own a PlayStation Portable (PSP)? If so, you can share media files between your PSP and your Xbox 360. Simply plug your PSP into the 360 using a USB cable, and presto—instant cross-platform file sharing. The Xbox 360 works with any type of portable device. It does not discriminate based on brand thanks to the nature of the Universal Serial Bus (USB). See Figure 2.31. To learn more interesting things you can do with your PSP, be sure to check out the book *The Gadget Geek's Guide to Your Sony PlayStation Portable* (ISBN 1598632361) by Jerri Ledford.

Figure 2.31
Memory Stick Pro Duo
provides massive storage
for the PSP, which is a
bridge to the Xbox 360
via USB.

Viewing Photos on a Data CD or DVD

You can insert a standard data CD or DVD packed with photos into the Xbox 360's DVD drive and then browse the photos on the disc like the portable device browser. When you insert a data disc, the Current Disc option becomes available, as shown in Figure 2.32.

Selecting the Current Disc option brings up the photos on the CD or DVD you've inserted into the 360. Figure 2.33 shows a CD loaded with photos from the Hot August Nights 2003 car show.

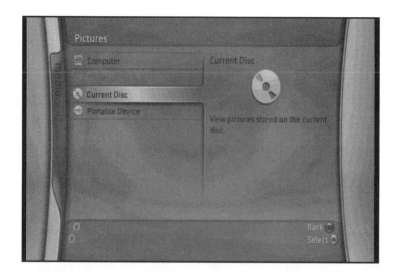

Figure 2.32
The photo browser is capable of reading a data CD or DVD loaded with digital photos.

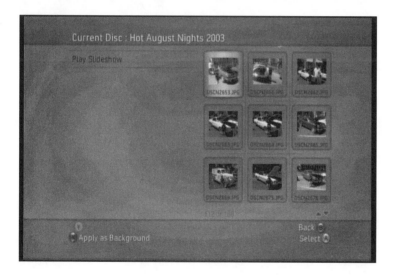

Figure 2.33
Viewing a photo slideshow from a data CD or DVD.

Viewing Photos on Your Digital Camera

Your Xbox 360 also supports digital camera interfaces via USB, but only if the camera is Plug-N-Play. Some cameras require a driver to be installed before they are recognized, which means that the electronics in such cameras do not have true Plug-N-Play support built in. The basic difference here is between devices that have *passive* file systems versus devices that have *active* file systems. A passive connection provides a bridge to files stored on it. An active connection provides USB connectivity in the device itself so that it is platform independent, able to work with any USB port—on Windows, Mac, Linux, and on consoles such as the Xbox 360 that are so equipped. Figure 2.34 shows a digital camera that has been plugged into the 360 using a USB cable.

Figure 2.34
Your Xbox 360 should recognize any digital camcorder or camera with a USB interface.

When you connect a digital camera, the Pictures screen enables the Digital Camera option, allowing you to select it (see Figure 2.35). Selecting this option brings up the Digital Camera

Figure 2.35
The Digital Camera option becomes available when you plug in your camera via USB.

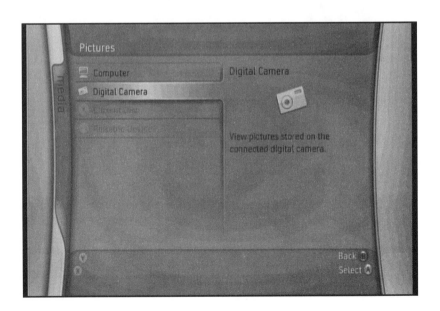

screen and provides access to the file system of the camera, with thumbnail views of the photos stored on the camera's memory card. (In the example in Figure 2.36, I'm using an SD, or Secure Digital, card.) From this point, you can select individual photos to view or run the slideshow to view all the photos in your camera.

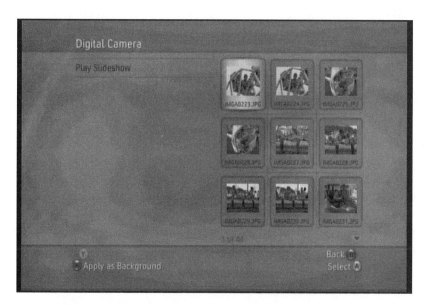

Figure 2.36
The thumbnail photo browser for a digital camera functions the same as the browser for other devices.

PLAYING VIDEO WITH YOUR XBOX 360

Unlike the media options for working with digital photos and music files, the Xbox 360 does not have extensive capabilities for working with video files. For one thing, there is no option to play a video file (such as an MPG or WMV) from a portable device or data CD/DVD. Figure 2.37 shows the Video option in the Media folder.

The only option available in the Video screen, as you can see in Figure 2.38, is to play videos found on the hard drive. If your 360 is not equipped with a hard drive, the Video option will not even be available. Selecting a video from the list starts the playback of that video (see Figure 2.39).

Figure 2.37
The Video option in the
Media folder menu.

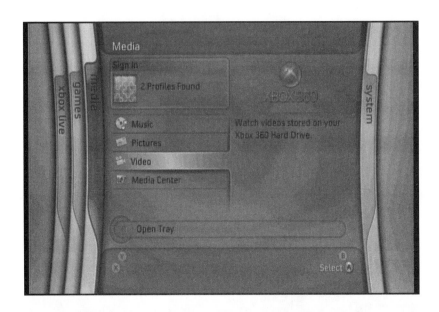

Figure 2.38
The list of videos stored on
the hard drive.

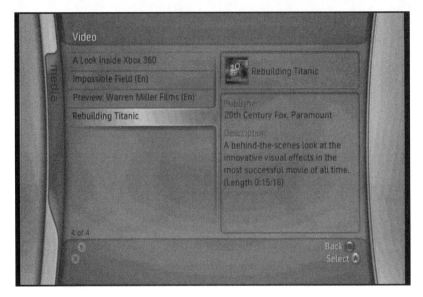

Figure 2.39
The video playback functionality is limited to pause and stop.

the gadget geek's guide to
Your Xbox 360

3
Xbox 360 Games

Your Xbox 360 is, without a doubt, the most powerful video game console system ever created. Although the multimedia capabilities that we explored in the previous chapter are interesting, the true power of this system is revealed only when you insert a game disc and play a next-generation game on this system. In this chapter, I am going to take you on a tour of some impressive games designed for the 360. We'll also get into the subject of backward compatibility—that is, which games from the previous Xbox you can play on your 360.

FEATURED GAMES

I'd like to explore three popular Xbox 360 games with you, because these games are representative of the launch titles available when the Xbox 360 was first released. Given the high quality of these first-generation games, I am looking forward to the upcoming second- and third-generation games for this console!

Project Gotham Racing 3

Project Gotham Racing 3 follows in the footsteps of two extraordinary predecessors that set new benchmarks for racing game realism. The first *Project Gotham* game was popular because it boasted a realistic driving experience that simulated actual conditions of the road. You could dent, damage, and wreck your car if you were not careful. *Gotham 2* brought even more detail to the franchise, adding more detailed levels and cars, better automobile physics, and a larger number of vehicles to drive. *Gotham 2* was one of the most technically advanced Xbox games that truly pushed the console to its limits. (And the occasional frame rate drop was a testament to the high level at which this game pushed the hardware.)

Game Features

Let's look at the features of this game and then jump in and drive solo and online. Figure 3.1 shows the beautifully designed cover art for the retail game box.

If you look at the back of the DVD case at the game specifications (see Figure 3.2), you'll note the following standard features:

- Players 1–2
- System link 2–8
- 350KB to save game
- HDTV 720p

These features describe the overall facets of the game to let you know that the game runs at the standard HDTV 720p, supports system link (for a LAN party!), and so on. Beneath the "green" specification boxes are some red boxes that describe the online features of the game:

Figure 3.1
Project Gotham Racing 3 is an extraordinary game for a launch title.

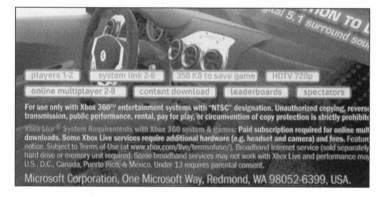

Figure 3.2
Features of the game.

- Online multiplayer 2–8
- Content download
- Leaderboards
- Spectators

Each game typically lists its own custom set of features on the back of the retail packaging, but most game publishers stick to the standard descriptions to make it easier for consumers to determine the capabilities of a game.

Playing Solo

The single-player game uses the same Gamertag as the online game, so you can play PGR3 offline to unlock cars, unlock new race tracks, and improve your score in the game. These features are available in the online game. Figure 3.3 shows the title screen.

Figure 3.3
The title screen of PGR3.

The first option, Gotham Career, allows you to play the game using your Gamertag either offline or online. The second option, Playtime, allows you to quickly jump into a race using any of the cars in the game (where the career mode requires you to purchase a car before you can drive it). The third option, Gotham TV, allows you to watch other players race online as a spectator. The fourth option, Achievements, lets you view your awards earned in the game. The last option, More, takes you to an options screen where you can configure the controller, choose a different Gamertag profile, and look for online updates to the game.

Selecting Gotham Career brings up the screen shown in Figure 3.4. From this screen, you can choose to play the game in solo or in online career mode. Either way, your credits (used to purchase new cars), kudos (used to unlock cars and tracks), and score are shared, regardless of whether you are playing online or offline. Because you are limited to owning only four cars at a time, it is a challenge to decide which ones to keep or sell. The type of cars you own determines what types of races you'll be able to enter, because there are classes (A to E) based on vehicle performance.

None of the features means a thing if the gameplay and racing physics are not top notch—and in this department, PGR3 delivers! Figure 3.5 shows an interesting race: a 1967 Shelby GT-500

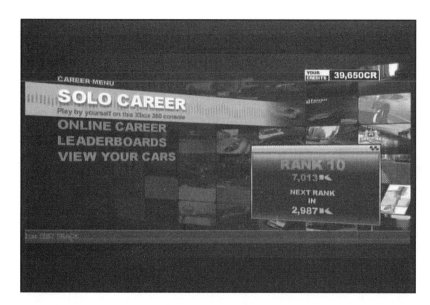

Figure 3.4
The Career menu in
PGR3.

and a 2007 Shelby GT-500, head-to-head. When you crash your car, your car will show actual damage, including body panels that wrinkle on the sides, front, and rear, and mirrors that break off. The hood or deck lid may even come loose.

Figure 3.5
The vehicle and driving
physics in PGR3 are top
notch.

Looking for Downloadable Content

Selecting More and then choosing Xbox Live Marketplace takes you to the PGR3 download screen, showing the available content for the game. As you can see in Figure 3.6, at this time there is no downloadable content for PGR3.

Figure 3.6

Viewing the available downloadable content for PGR3 on Xbox Live.

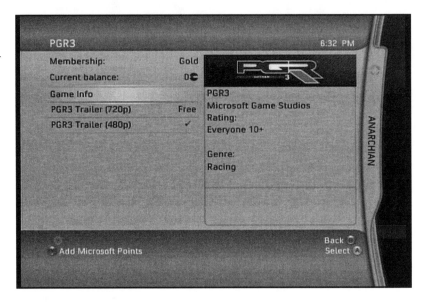

Gotham TV: Spectator Mode

The Gotham TV is exciting enough to become a standard feature in many other games. When you select this option, you are presented with the Gotham TV menu shown in Figure 3.7. From this menu, you can get a list of online races taking place now and then open that channel to watch the race live!

Figure 3.8 shows a live broadcast from Gotham TV of an actual online race taking place. When you're watching Gotham TV, you can change the camera view using the LB button.

Figure 3.9 shows another Gotham TV broadcast, featuring another top leaderboard racer. You can even view the game through the eyes of the player from inside the car, as you can see in Figure 3.10.

Kameo: Elements of Power

Rare Studio was once a famous game development studio that produced top-notch games for Nintendo, dating back to the infamous *Donkey Kong Country* for the Super NES. Now Rare has found a home within the halls of Microsoft, producing games for the Xbox as an exclusive

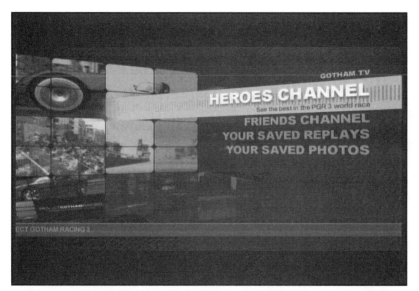

Figure 3.7
The Gotham TV options menu.

Figure 3.8
Viewing a live "Gotham TV" broadcast of an actual online race.

in-house developer. This studio has truly shown its talent with *Kameo*, which is a remarkably original, creative work of art. Varied gameplay requires the gamer to think creatively to solve difficult problems using one of three characters that *Kameo* can morph into.

Figure 3.9
Another top racer show-
ing off for Gotham TV.

Figure 3.10
Viewing the race from the
driver's perspective.

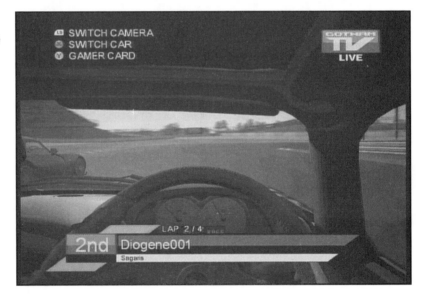

Interactive Work of Art

Refer to Figure 3.11, which is another beautiful example of highly detailed retail box art. You
can see that *Kameo* is a single-player game with a cooperative feature for a second player, with
support for the usual HDTV 720p, content download, and leaderboards.

Figure 3.11
Kameo: Elements of Power was created by Rare.

This game is a traditional single-player adventure game with challenging levels, combat with a wide variety of creatures, and interesting puzzles to solve. The graphics in this game are stunningly beautiful (see Figure 3.12), reminding me of another gorgeous game, Peter Molyneux's *Fable*.

Figure 3.12
The graphics in Kameo are beautiful, with animation and special effects that cannot be properly captured in print.

The goal of *Kameo* is to make your way through a series of gigantic levels, each of which is divided into smaller areas that require some action to move to the next part of the level. The main character's name is Kameo, as shown in Figure 3.13.

Figure 3.13
Kameo is the main character of the game.

Characters

Kameo can morph into many different types of creatures, and the three creatures you can use at a time will change to new creatures as the game progresses. One of the most interesting characters that you start out with is Chilla, who can throw ice shards at enemies (see Figure 3.14). In addition to these three characters, Kameo gets to use many more characters throughout the game with special abilities.

Another fascinating character that Kameo can control is Pummel Weed, shown in Figure 3.15. Pummel Weed is a boxer who knocks the lights out of bad guys.

The third character that you can morph into during the early parts of the game is Major Ruin (shown in Figure 3.16). This character can spin and roll around, knocking enemy characters off their feet.

Perfect Dark Zero

Perfect Dark was sort of a derivative game created by Rare using the *Goldeneye 007* engine for the Nintendo 64. *Goldeneye* was one of the most highly acclaimed games of 1998, with an engaging story and immersive gameplay that was downright fun to play and included the great

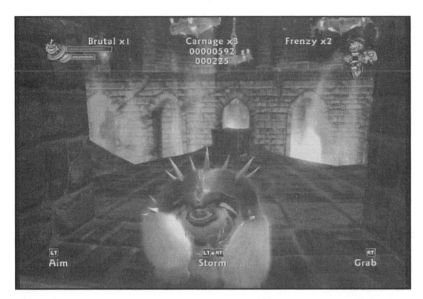

Figure 3.14
Chilla, one of the characters that Kameo can change into, is good at hand-to-hand combat.

Figure 3.15
Pummel Weed is another interesting character who's good at boxing and sliding under doors.

four-way split screen "deathmatch" gameplay. Along with *Mario 64*, *Goldeneye* is the other remarkable game I recall playing on the Nintendo 64. (Although there were many other great games for this platform, these are among the most memorable.)

Figure 3.16
Major Ruin is an interesting character who can roll around to bash into enemies.

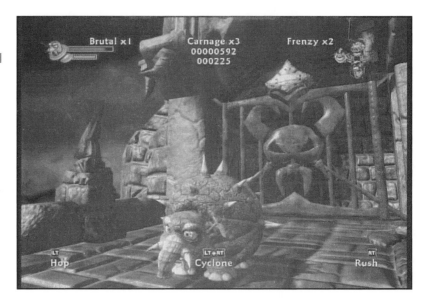

Rare followed up with *Perfect Dark* a couple years after the release of *Goldeneye* using an enhanced version of the *Goldeneye* engine (an engine that might have pushed the hardware a little too much). As Figure 3.17 shows, *Perfect Dark Zero* also supports four-player split-screen gameplay, with support for cooperative two-player mode and online multiplayer gameplay via Xbox Live.

Figure 3.17
Perfect Dark Zero is a mission-based first-person shooter.

The single-player experience is mission based, meaning that you must follow a certain path and complete specific objectives to continue. The action is fast paced but can be a bit too easy for an experienced gamer, who will want to play the game at a high difficulty level to be challenged. Figure 3.18 shows the gameplay in an online multiplayer team deathmatch.

Figure 3.18
The online multiplayer feature of the game will continue to entertain after you've completed the single-player campaign.

PLAYING ORIGINAL XBOX GAMES ON YOUR 360

At the time of this writing (December 2005), the Xbox 360 supports 204 original Xbox games, with support for additional games coming every month. This is a significant feat of engineering because the Xbox 360 is as foreign in architecture to the original Xbox as a PC is to a Mac, with completely different processors, graphics chips, and instruction sets.

The most significant disclaimer you need to be aware of regarding backward compatibility is that your Xbox 360 must be equipped with a hard drive (shown in Figure 3.19) to play original Xbox games on it. This is an unavoidable requirement because the original Xbox came with a hard drive, and all original Xbox games utilized it. Because the Xbox 360 Premium package comes with the hard drive for only $100 more than the Core package, this is a savings of $50 if you purchase the hard drive separately. If you have the Core package, the hard drive will be an inevitable upgrade because it is indispensable.

Meltdown

The Xbox 360 hard drive is required for playing original Xbox games. There is no workaround for this requirement. For this reason, the hard drive accessory is an inevitable upgrade if you don't own one already.

Figure 3.19
The Xbox 360 hard drive comes with the Xbox 360 Premium package.

Meltdown

Most of the material in this section of the chapter assumes that you have Internet access and an Xbox Live account. For a more thorough tutorial on getting online with your Xbox 360, refer to the following chapters for more information: Chapter 5, "Connecting Your Xbox 360 to Your Home Network" and Chapter 7, "Going Online with Xbox Live"

Nearly every triple-A title will run on the Xbox 360—and that consists of online gameplay via Xbox Live, including the following games. (Note that this is just a partial list.)

- *Halo, Halo 2*
- *Star Wars: Knights of the Old Republic I & II*

- *Crimson Skies: High Road to Revenge*
- *Metal Arms: Glitch in the System*
- *Tom Clancy's Ghost Recon 1 & 2*
- *Tom Clancy's Rainbow Six 3*
- *Tom Clancy's Splinter Cell* series
- *Dead or Alive 3*
- *Need For Speed Underground 2*
- *Fable*
- *Grand Theft Auto* series
- *Prince of Persia: The Sands of Time*
- *Half-Life 2*
- *Jade Empire 3*
- *Max Payne 1 & 2*

Thankfully, *Barbie Horse Adventures: Wild Horse Rescue* also runs on the 360. The complete list of supported games is updated regularly at http://www.xbox.com/en-US/games/backwardcompatibilitygameslist.htm.

Given that these two platforms are so different, how do you suppose Microsoft has pulled this off and made it possible to play original Xbox games on the 360? Well, every original Xbox game requires a download from Xbox Live before it will run. To save time, all the individual game patches are bundled together into a single download. The result is that each game is emulated on the Xbox 360.

Beyond the Manual

There is a Web page at http://www.xbox.com that explains how you can download and install the Original Xbox Game Support update for your Xbox 360 (if you are not an Xbox Live subscriber). The process involves downloading a file to your PC, extracting it from the zip archive, and burning that file to a blank CD-R, which the Xbox 360 can load and run. The URL for this update page is http://www.xbox.com/en-US/games/backwardscompatibility.htm.

The original Xbox and the Xbox 360 have similar development kits that game publishers use to create games for these systems. If you also play Windows PC games, you are probably familiar with Microsoft's DirectX game library, which is the primary library used for PC games today. Even games that use a cross-platform graphics library such as OpenGL usually use DirectX components for device input (to support a keyboard, mouse, or joystick), sound effects, and music. id Software's games are known for being cross-platform friendly, meaning that their games (such as the popular *Quake* and *DOOM* series) are compiled to run on Windows, Mac, and Linux PCs. Many game developers do not support multiple platforms, but id Software has been a proponent of this technology for many years. In fact, that is the reason why *DOOM III* and *Quake 4* are already available for consoles.

This same DirectX library is also used in the development of Xbox and Xbox 360 games. Although the 360 is far more powerful and capable, games for the 360 are programmed in a similar fashion to games developed for Xbox. If I might use an analogy, the Xbox is to the Xbox 360 like the Ford Mustang is to the Ford GT. These cars differ dramatically in performance, but they are based on the same technologies and are built in the same factories. *Project Gotham Racing 3*, one of the games featured earlier in this chapter, provides an extensive list of production cars that you can drive, such as the cars shown in Figure 3.20.

Figure 3.20
*Project Gotham Racing 3
allows you to own up to
four cars at a time.*

Beyond the Manual

To see the comparison for yourself, the Ford GT's Web site is at http://www.fordvehicles.com/fordgt, whereas the Mustang site is at http://www.fordvehicles.com/cars/mustang.

Installing the Original Xbox Games Support Update

Even if you don't have access to Xbox Live, you can update your Xbox 360 to support original Xbox games by downloading the update from http://www.xbox.com/en-US/games/ backwardscompatibility.htm and following the instructions on that page. You can also order the Original Xbox Games Support CD from a link on that page.

I'm going to walk you through the process of running an original Xbox game on the 360, using *Crimson Skies: High Road to Revenge* as the example game. As soon as I insert the *Crimson Skies* disc, the Xbox 360 displays the screen shown in Figure 3.21.

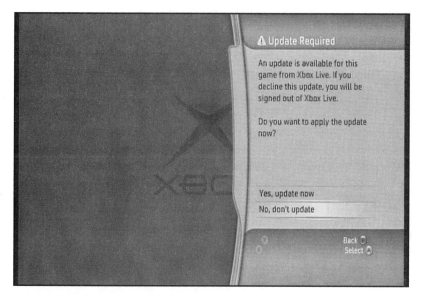

Figure 3.21
Crimson Skies: High Road to Revenge is a fun arcade-style flight simulator game.

If you don't have access to Xbox Live (or if your 360 is not jacked into the network), the screen shown in Figure 3.22 is displayed instead, explaining that the emulator patch for the game is not available. This screen is also displayed when you try to play an original Xbox game that is not supported on the 360.

Selecting Yes, Update Now brings up the update screen shown in Figure 3.23.

Figure 3.22
The Original Xbox Games Support patch is downloaded and installed.

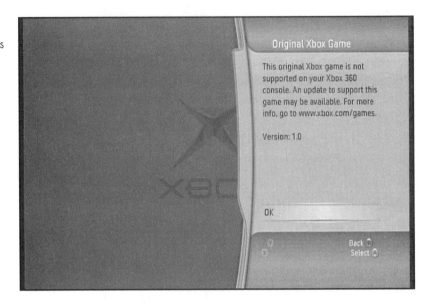

Figure 3.23
The Original Xbox Games Support patch is downloaded and installed.

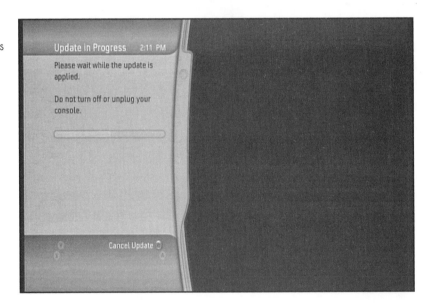

After the update has been downloaded and installed, you are returned to the Xbox 360 Dashboard. That's because the update installer does not automatically run the game afterward.

You can move the selection to the Play Game option at the bottom of the screen, shown in Figure 3.24. You can also eject and reinsert the disc in the DVD drive to launch the game.

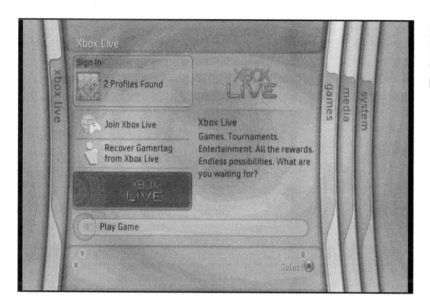

Figure 3.24
Preparing to run *Crimson Skies* after the emulation patch has been installed.

Playing *Crimson Skies* on the 360

What happens next is very interesting. When you relaunch the game (*Crimson Skies*, in this case), the Xbox 360 actually runs an emulator program that was installed with the Original Xbox Games Support update. This emulator program comes up first and displays the logo shown in Figure 3.25 before running the game.

Next, the game title (or loading) screen comes up, which means the game is successfully running on your 360 (see Figure 3.26). This occurs for every original Xbox game you insert into your 360 to play, as long as it is supported by the current set of emulator patches. Some original games will not play on your 360—at least not until the Xbox team creates the emulator patch for all of them. Because the Xbox 360 has only been available for a month (at the time of this writing) and there are already more than 200 games supported, I think it's safe to say that every original Xbox game will eventually play on the 360.

As you can see in Figure 3.27, *Crimson Skies* runs perfectly on the Xbox 360. There is a difference in the gameplay, though. Because of the 360's powerful processor, games that previously pushed the limits of the Xbox run more smoothly on the 360. For instance, I have noticed that every once in a while, *Halo 2* would slow down a bit when playing an intense multiplayer network game at a LAN party, especially when a lot of players are on the screen at once. But there is never a slowdown when playing such games on the 360.

Figure 3.25

An emulator program kicks in to run original Xbox games.

Figure 3.26

The original Xbox game is now running on the 360 via emulation.

How does the emulator work? Are you familiar with the Microsoft Virtual PC program for Mac OS systems? Well, because the Xbox 360 is already based on DirectX and a highly modified version of the Windows core operating system, the original Xbox game emulator is not nearly as extensive as Virtual PC for Mac. Instead, it is more likely that a simpler function call translator or "bridge" is used to adapt Xbox code (compiled for an Intel Celeron 733MHz processor) into

Figure 3.27
Crimson Skies: High Road to Revenge runs flawlessly on the Xbox 360.

Xbox 360 code (that will execute on the triple-core Power PC 3.2GHz processor). It's not true emulation; it's more like on-the-fly native code *translation*, with most of the underlying features being relied upon (such as TCP/IP networking used to connect to Xbox Live).

Problematic Emulation

Unfortunately, not every original game runs as well on the 360 as *Crimson Skies* (which is probably true of all the top-selling games). Lesser-known niche games such as *Star Trek: Shattered Universe* do not play quite as well. Although the menu is somewhat usable in this game (see Figure 3.28), the actual gameplay is totally messed up (see Figure 3.29).

Because the cinematic video at the beginning of *Shattered Universe* is rendered in an unusual scale with the right edge of the screen left blank, I might assume that the game uses some unconventional graphics routines to display the frames of a video sequence—and this would have a diverse effect on the entire game. However, I decided to check the video settings of my 360, and it turns out that *Shattered Universe* supports HDTV 480p—a sure sign that the problem here is related to the display settings, not the game. The screen is actually messed up in this game because the 360 is trying to resample the video on the fly from widescreen to normal display mode. There is a solution to this problem.

Go into the System folder on the Dashboard and open the Console Settings screen, shown in Figure 3.30. There's the problem right there. The display is set to standard TV, but the mode is set to widescreen. Because *Shattered Universe* supports HDTV 480p—not 720p—changing the display setting from widescreen to normal should fix this game.

Figure 3.28
Some original Xbox games do not play well on the 360.

Figure 3.29
What could cause a game like *Star Trek: Shattered Universe* to render so badly?

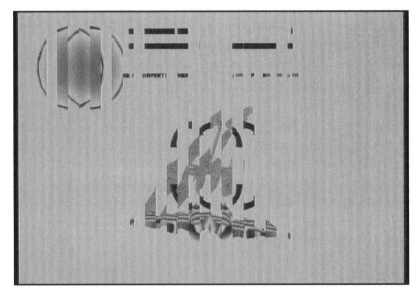

Within the Display screen are two options: HDTV Settings and Screen Format. Select the Screen Format option to change the screen to normal, as shown in Figure 3.31.

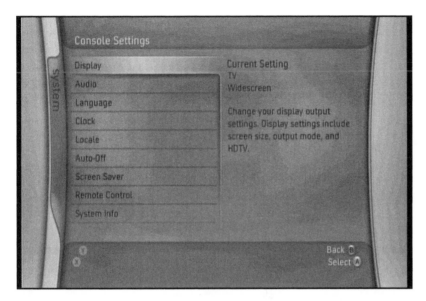

Figure 3.30
Changing the display settings.

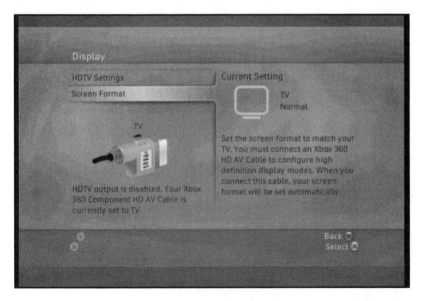

Figure 3.31
The display is now set to normal resolution (not widescreen).

Firing up *Shattered Universe* again results in a dramatic improvement. Figure 3.32 shows the same title screen that you saw in Figure 3.28, only now it is displayed without distortion. When you're inside the game, the graphics look exactly as they should, without distortion (see Figure 3.33).

If you have an original Xbox game that you are having trouble playing on your 360, and you found that game in the list of supported games, it is possible that the problem lies within your

Figure 3.32

The title screen of *Shattered Universe* now looks normal.

Figure 3.33

The in-game graphics engine is now rendering correctly.

360's display settings, not with the game. This is especially true if you own a high-definition television (HDTV) and enjoy playing games at 720p or 1080i; there are many original Xbox games that do not run in HDTV. There is a convenient solution to this problem built into the High-Def cable, which comes with the standard composite video cable and the HDTV component cables. If you have both sets of cables plugged into the appropriate inputs on your TV (assuming it has a composite RCA-style input), you can simply flip the video switch on your HDTV cable to change it from HDTV to RCA. (Take a look at the switch diagram in Figure 3.31.) If you purchased the Xbox 360 Premium package, you already have the HDTV component cable, but the Core package only comes with a composite (RCA) cable.

Playing *Ford Mustang* on the 360

Another good test of backward compatibility on the 360 is the game *Ford Mustang*, which came out in 2005 and is a niche game (something that perhaps only a Ford fan would enjoy, although any casual racing fan will appreciate the playability of this game). It is far more likely that older niche games will be added to the compatibility list rather than newer games, which is why *Ford Mustang* is a good test case.

After you insert the disc into the 360, the title screen comes up (see Figure 3.34)—so far, so good. Navigation among the in-game menus appears to function normally, including the selection of a car to drive (see Figure 3.35).

Figure 3.34
The title screen of *Ford Mustang*.

Figure 3.35

Selecting a Mustang to race in the game.

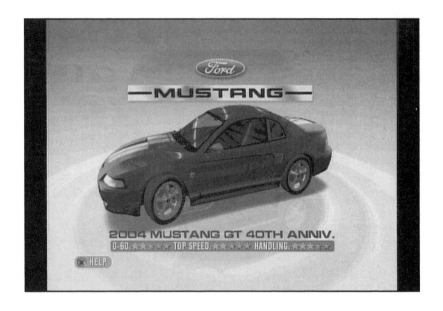

The real test of whether a game will run, of course, is watching the actual gameplay using the game's 3D graphics engine. As you can see in Figure 3.36, the *Ford Mustang* game is being rendered as it should, and it looks terrific on the 360. In fact, this game is more forgiving than *Shattered Universe*, because it runs fine in either normal or widescreen TV mode.

Figure 3.36

The *Ford Mustang* game engine looks exactly as it should.

Playing Original Xbox Games on Live

You will be happy to know that original Xbox game emulation works as flawlessly online as it does for single-player games offline, with full Xbox Live support. As is the case with the original Xbox Live, most online games require a patch to update them to the latest version. Playing games online tends to require more out of a game than a single-player experience does, so flaws in the game are more evident when playing online. In addition, game developers often release new versions of a game to improve the online gameplay, and they often release new levels and upgrades to the game that you can access on Live.

The previous step to download a patch for playing original Xbox games on the 360 is separate and distinct from any updates that may be required to play on Live from within the game. Most Live games released during the three years that Xbox Live was operating before the release of the 360 have been updated to improve gameplay and provide new features. *Crimson Skies* is one such game, so it required an update as soon as I tried to log into Live, as shown in Figure 3.37.

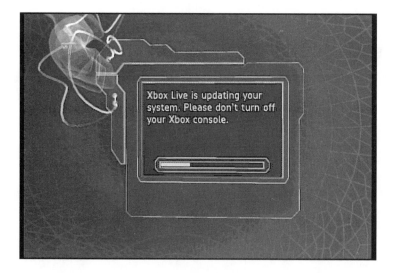

Figure 3.37
Most original Xbox games must be updated for online play on Xbox Live.

After you're connected online, the Xbox Live menu of *Crimson Skies* notes that some new content is available for download (see Figure 3.38). This is yet a third instance of downloading we've come across while preparing to play this original Xbox game. However, everything you see inside the game is normal for that game and operates in the same manner on your original Xbox at this point. Only the first download (for the emulator patch) was different on the 360.

Opening the downloadable content screen (shown in Figure 3.39) in *Crimson Skies* reveals that numerous updates are available for the game—and they are all free updates.

Figure 3.38
New downloadable con-
tent is available for this
game.

Figure 3.39
The list of new updates for
the game.

Here is what is available for download:

- Three new levels
- One new game type
- Four new aircraft

Selecting one of the available content items downloads the item to your Xbox 360. In Figure 3.40, you can see that I have chosen to download the Vampire airplane to add this aircraft to my game. I recommend downloading all available content for a game because most online multiplayer games require the latest updates.

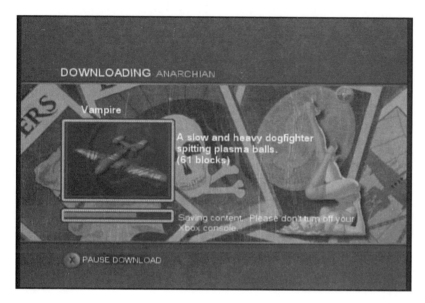

Figure 3.40
Downloading a new content item for *Crimson Skies*: the Vampire airplane.

4

Gadget Geek Gear: Xbox 360 Accessories

The earliest video game consoles typically included two controllers and offered little or no additional accessories. Although the first NES came with a goofy "robot" for the North American release, the motive for it was to market the NES as an educational tool because Nintendo believed that consumers would not buy a video game console purely for entertainment. Times have changed! Accessories are all the rage today for the latest consoles, and consumers are buying them. It was once true that most consumers never bought additional hardware accessories for their console and used only what came in the retail box. But today, with so many different types of televisions (from HDTV to S-Video to composite) and so many genres of games, accessories such as high-definition cables, wireless controllers, and driving wheels are popular.

CONTROLLERS

The Xbox 360 is the first video game console to offer a wireless controller as standard equipment (in the Premium package). The Core package comes with a wired controller.

Standard Controller

The standard wired controller is shown in Figure 4.1. This controller has been refined quite a bit since the Controller S on the original Xbox came out—greatly improving on the original Xbox controller, which was bulky and uncomfortable to use in the opinion of many.

Figure 4.1
The Xbox 360 standard wired controller.

Wireless Controller

The wireless controller is shown in Figure 4.2. This controller uses a radio frequency transceiver built into the Xbox 360 (located just behind the power button).

Figure 4.2
The Xbox 360 wireless controller provides excellent and precise responses to input.

The wireless controller is powered by two AA batteries or a rechargeable battery pack that is optionally available. The Charge & Play kit, shown in Figure 4.3, comes with a rechargeable battery. A controller cable (USB, like the wired controller) plugs into the small socket on the front edge of the wireless controller, and the other end plugs into one of the USB ports on the bottom front of the Xbox 360 where you normally plug in the wired controller. This allows you to continue playing games while the battery pack is recharging, and the nice thing about it is that the recharger draws its power from the USB port.

Figure 4.3
The Charge & Play kit for the wireless controller.

Voice Communication Headset

The voice communication headset (see Figure 4.4) plugs into the back of the Xbox 360 controller facing the player and provides voice chat capability to your games. The nice thing about the headset is that it works even with the wireless controller!

Figure 4.4
The voice communication headset plugs into the controller.

If you are a racing fan and would like better control in games like *Project Gotham Racing 3*, I recommend you pick up a racing wheel from a company such as Mad Catz (http://www.madcatzstore.com), whose products are usually available at major retailers.

REMOTE CONTROLS

Two remote controls are available for the Xbox 360: the Premium Media Remote and the Universal Media Remote. Both remote controls add a lot of functionality to your Xbox 360 and are sometimes more convenient for doing certain tasks in the Dashboard.

Premium Media Remote

The Premium Media Remote (see Figure 4.5) is bundled with the Xbox 360 Premium package and is not available for purchase separately (at least not at the time of this writing).

Universal Media Remote

The Universal Media Remote (see Figure 4.6) is the remote control that is available in retail outlets for your Xbox 360. This remote is actually more functional than the smaller one. It provides some interesting control over the 360 when you are connected over the network to a

Windows Media Center PC, in which case you can stream live TV through your Xbox 360 and perform tasks like recording and pausing the video stream.

Figure 4.5
The Premium Media Remote.

Figure 4.6
The Universal Media
Remote.

STORAGE DEVICES

There are two options for storing data for use in your Xbox 360: the memory unit and hard drive.

Memory Unit

The memory unit for the Xbox 360 (shown in Figure 4.7) is significantly denser than memory cards in other consoles to date, providing 64MB of storage space for saved games. You can also use the memory unit to transport your Xbox Live account to another Xbox 360 when you would like to play at a friend's house.

Figure 4.7
The memory units can store 64MB of saved games and your Xbox Live account.

Figure 4.8
The standard hard drive has 20GB of storage space.

Hard Drive

The Xbox 360 hard drive (shown in Figure 4.8) is standard equipment in the Premium package but does not come with the Core package. The hard drive is available for purchase separately if you don't already have one. (It is strongly recommended that you use one for the best possible experience with your Xbox 360.) If you play online using Xbox Live, it's possible to go without a hard drive, but if you want to download additional content (such as new levels and items for your games), you need the hard drive. In addition, if you want to play original Xbox games, the hard drive is required. Although the hard drive is currently only 20GB, it is likely that new, higher density hard drives will be available for the Xbox 360 in the future.

AUDIO/VIDEO CABLES

Audio and video cables are often the most confusing consideration when you are thinking about buying a new video game console system. I will try to clear this matter up for you with regard to the Xbox 360 to help you select the best cable for your needs.

Composite "RCA" Video Cable

The composite or RCA-style cable (see Figure 4.9) comes standard in the Xbox 360 Core package and provides support for standard television sets. This cable is what you will want to use if you have a standard TV (which is known as SDTV). But if you have a high-definition television (HDTV), you will want the component cable, described next. Figure 4.9 shows the composite

Figure 4.9
The composite cable works with standard television sets.

cable. Standard TV sets have a maximum resolution of 640×480 and are either interlaced (i) or progressive (p). However, not all video sources output at this resolution. Table 4.1 illustrates.

Table 4.1 SDTV Video Resolutions

Type	Ratio	Resolution
Broadcast TV	4:3	440×330
VHS Tape	4:3	320×240
LaserDisc	4:3	560×420
DVD	4:3	640×480

If you've ever wondered why VHS movies look so terrible compared to DVD video, this table shows why. At a resolution of 320×240, how did this format ever become so popular over technologies like LaserDisc? Although an attempt was made to produce an optical disc called Super VHS using existing CD-ROM technology that boasted a 1:1 resolution of 480×480, this format never caught on.

In the TV industry, it is standard to list the vertical resolution (such as 330 lines for broadcast TV); however, the horizontal resolution is never fixed and varies from one TV set to another, which affects quality. SDTV refreshes the display at 30 frames per second (fps). If you have a standard DVD player, it outputs video at 480i. If you have a progressive-scan DVD player, it outputs video at 480p. (And the difference is dramatic.) The Xbox 360 SDTV cable outputs a 480p video signal.

Beyond the Manual

There is a misconception about 480p that many gamers have. 480p is not HDTV; it is SDTV. HDTV operates at 720p, 1080i, and 1080p. Video games running at 480p look much better due to the progressive scan of the video, but only if the TV set is capable of progressive display. Most standard TVs are not.

S-Video Cable

A related technology, S-Video uses the same basic formats as composite video, but it separates the color signal (chrominance) from the brightness signal (luminance), which produces a much cleaner, sharper image than is possible with composite. S-Video greatly improves the image, especially in video games, because it eliminates problems like color bleeding, which tends to decrease the quality of a composite display. The Xbox 360 S-Video cable is also shown in Figure 4.9 (as a combined set). If your TV set is not HDTV, your best option is to check to see if it has an S-Video input and then purchase the S-Video adapter for your Xbox 360. Aftermarket S-Video and composite wires are usually combined on the same cable.

Geek Speak

Chrominance refers to the color portion of a video signal. *Luminance* refers to the brightness of a video signal. S-Video separates the two (for a clearer video signal), whereas standard composite video combines them.

Component HDTV Cable

The cable to use for a High-Definition Television (HDTV) is the component video cable (shown in Figure 4.10). The word *component* describes how this cable works. It uses three cords for video output to an HDTV set: usually a red cable, a green cable, and a blue cable, as you can see in Figure 4.10. These three cables are bundled together and collectively called the component cables, representing the Y/Pb/Pr (YUV) inputs for HDTV. Table 4.2 illustrates.

Figure 4.10
The component video cable carries the high-definition signals to your HDTV.

Table 4.2   HDTV Video Resolutions

Mode	Ratio	Resolution	Frame Rate
480i	4:3 or 16:9	704×480	30 fps
480p	4:3 or 16:9	704×480	60 fps
720p	16:9	1280×720	60 fps
1080i	16:9	1920×1080	30 fps
1080p	16:9	1920×1080	60 fps

VGA Video Cable

The Xbox 360 VGA HD cable (shown in Figure 4.11) has a DV-15S VGA connector for sending the video signal from your Xbox 360 to a computer monitor. Although this type of cable won't work with a digital monitor (where a DVI-D cable is needed), the VGA cable is a nice accessory if you have a high-quality monitor available or if you just want to play with your 360 without requiring a TV (using your computer monitor). This cable does not output at HDTV resolutions, with a maximum supported resolution of 1280×1024 (SXGA, 5:4 ratio) and 1360×768 (WXGA, 16:9 ratio).

Figure 4.11
You can use the VGA adapter cable to send your Xbox 360's video signal to a computer monitor.

NETWORKING

The Xbox 360 comes with a built-in network adapter, so all you have to do is plug in a Local-Area Network (LAN) cable from the 360 to your router or hub and you are good to go. However, what if your 360 is located in an inconvenient place that is nowhere near the router or hub (or your broadband adapter)? In that case, rather than routing a long cable through your house, the best option is to use the wireless adapter, shown in Figure 4.12. Of course, to use the wireless adapter, it goes without saying that you also need a wireless access point already set up.

Beyond the Manual

I cover networking in far more detail in the next chapter, so I'm only briefly touching on the subject here.

If you don't have a network router or hub yet, and you already use your broadband adapter for your PC, you will need a router or hub to share the broadband connection between your PC and Xbox 360. This is the job of a router. You can now buy a router that has a four-port network switch and wireless access point combined. They are affordable today, typically costing less than $50 at major electronics stores. (I recommend http://www.newegg.com.)

Figure 4.12
The wireless network
adapter for the Xbox 360.

COSMETIC ACCESSORIES

You can customize your Xbox 360 by replacing the front cover with a custom-designed front cover of your choice. This offers you the opportunity to really make your 360 suit your personal style. As you can see in Figure 4.13, there are many different styles of front covers available, like

Figure 4.13
Dozens of designer
faceplates are available
for the Xbox 360.

these four samples. There are already several dozen available from various retailers online (such as newegg.com), and the list is bound to grow in time. Expect to see custom face plates available with themes from your favorite games and from themes in popular culture in the near future.

the gadget geek's guide to Your Xbox 360

5

Connecting Your Xbox 360 to Your Home Network

The Xbox 360 was designed from the ground up to be a connected video game system with significant networking capabilities out of the box. Like its predecessor, the Xbox, the 360 comes equipped with a built-in Local Area Network (LAN) port; but unlike the Xbox that is limited to 10 megabits (Mb), the 360 is capable of networking at 100Mb. I will explain the networking capabilities of the Xbox 360 in this chapter and show you how to configure your Xbox 360 to use the wireless adapter as a viable option.

Geek Speak

A *megabit* is one million bits. In the context of networking, this represents one million bits of transfer speed (or bandwidth) per second.

WHAT IS A NETWORK?

There is an underlying technology that the Xbox 360, your PC, and many other devices rely upon. This technology has evolved over the past three decades from very early, primitive modems (the earliest of which transmitted data at 110 bits per second) to the current networking standards of 100 megabit Ethernet and 1000 megabit (or gigabit) Ethernet. In recent years, the new technology of wireless Ethernet has been developed from its earliest roots with proprietary wireless technologies developed by individual companies to the recent wireless standards. An organization called the Institute of Electrical and Electronics Engineers (IEEE), composed of engineers and scientists in the computer and electronics industries, defines the standards that all companies must conform to in their products or risk market exclusion.

A *network* is an often misunderstood term. In its simplest definition, a network exists when two or more computers are connected by some means and are able to communicate by sharing data, files, or devices. For instance, when you plug your Windows XP system into your network router or switch and then connect another PC or device to that same router or switch, you have successfully created a two-node network. Connecting your Xbox 360 directly to your cable or DSL modem does not "network" your console—it is simply online. Having Internet access does not mean that your system is on a network. However, dialing in to your company's Virtual Private Network (VPN) *does* make your computer part of your work network; the difference is that you are dialed in remotely, and the telephone line is being used for the network connection. Similarly, browsing to a corporate Web site and logging into your computer at work (from your home PC) is another form of networking. But the act of browsing to amazon.com or ebay.com does not establish a network.

Why are IEEE standards so important? Imagine, if you will, what it would be like if every network equipment company defined its own standard for network signals. This is actually what the networking industry was like in the early days before companies in this industry got together to agree upon the standards, which were then defined and formalized by IEEE. There is a close relationship between IEEE and the companies that build consumer electronics devices (such as networking equipment), because most of the engineers at these companies are members of IEEE. This organization defined the IEEE 802 standard for computer networks. Many substandards are part of the overall 802 networking standard, shown in Table 5.1.

Table 5.1   IEEE 802 Network Standards

Standard	Description
IEEE 802.3	Standard for Ethernet networks
IEEE 802.5	Standard for token-ring networks
IEEE 802.6	Standard for Metropolitan Area Networks (MANs)
IEEE 802.11	Standard for 2-Mbps wireless networks
IEEE 802.11a	Standard for 54-Mbps wireless networks
IEEE 802.11b	Standard for 11-Mbps wireless networks (Wi-Fi)
IEEE 802.11g	Standard for 54-Mbps wireless networks

As networking standards became more solidified, network devices that were manufactured by different companies could interconnect and work together. In the past, you might have needed to stick with one particular brand for a company's entire networking infrastructure. Today, all the 10/100 network devices follow the IEEE 802.3 standard. There are some minor subversions of 802.3, such as 802.3z, which is a standard defined for Gigabit Ethernet that transmits data at 1000 Mbps. Gigabit networks transmit data at ten times the speed of a standard 100-Mbps network (which the Xbox 360 supports). Most home networks do not benefit from Gigabit Ethernet because that is overkill for a typical home network. Even a huge LAN party in your home with perhaps 30 players operates fine at 100 Mbps. Games are designed to work with far less bandwidth.

Beyond the Manual

If you would like to learn more about home networking, I recommend the book *Home Networking Solutions* (Thomson Course Technology; ISBN 1929685513) by Paul Heltzel.

NETWORK GEAR

Network gear—such as network cards, hubs, and switches—remained fairly expensive through the 1990s. It was only when Internet Service Providers (ISPs) began offering broadband Internet access that network gear started to drop in price to a more affordable range for the average home network enthusiast. The earliest cable and DSL modems were offered to customers for lease only, meaning that you paid a monthly premium to rent the broadband modem in addition to the broadband Internet charges. As more and more people signed up for broadband, increasing quantities of sales drove prices down to a level where the average consumer could afford to buy a cable or DSL modem rather than renting one. At one time, I rented a cable modem for $15 per month in addition to the $40 per month broadband fee.

Network Cards

To connect to a network, your computer (or other computing device) must have a network card or chip built in with a network port for the network cable. There are many different network cables to choose from, so it helps to know what you need in advance. The Xbox 360 comes with a standard 100Base-T network port and includes a standard twisted-pair network cable. What this means in layman's terms is that the Xbox 360 works with any typical 10/100 network router, switch, or hub.

Network cards are no longer common because most motherboard manufacturers (such as Asus, Abit, Gigabyte, and Intel) are providing network ports on the motherboard now. The network card shown in Figure 5.1 is a typical representation of a Linksys card; network cards from other manufacturers look similar. Figure 5.2 shows another type of PC network card. In this case, it is a wireless notebook card. To use this card, you need a wireless router, a wireless switch, or a wireless access point (AP) that is connected to your router or switch. Many routers and switches now include built-in wireless access points.

Figure 5.1
This Linksys 10/100 LAN card is cheap and works great.

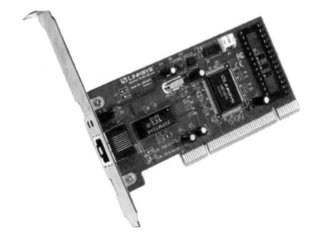

Figure 5.2
The Linksys 802.11g
notebook card.

Routers, Switches, and Hubs

Because wireless networking is so prevalent today, most networking gear supports wireless in one form or another. It's actually difficult to find wired-only routers and switches without built-in wireless support, but they are still available (in dwindling quantities). Competition is fierce in the network hardware industry. When one company is able to build a broadband router with a built-in 4- or 5-port switch *and* include a wireless access point—all in the same device—it's up to the other companies to follow suit with a similar product to remain competitive. The alternative is to purchase three separate devices.

First, you have your cable or DSL modem that is connected to the outside cable or phone line. Then you connect the broadband modem to a router using a standard network cable. That router acts like a bridge for your LAN, with hardware-based Network Address Translation (NAT) that makes it possible to play games on your PC or Xbox 360 online at high speed. Most routers include a built-in 4-port switch so that you can plug your devices directly into the router to give them broadband Internet access and simultaneously link them as a LAN. Some of the earlier routers provided just the bridge to your broadband modem, requiring a separate switch with four or more ports for your PCs. If you wanted to add wireless (802.11a, 802.11b, or 802.11g) support to your network, you had to purchase an access point and plug it into one of the free ports on your switch.

Things are much easier today. (Isn't it amazing how technology continues to drive forward, producing more powerful and complex devices that somehow seem to get easier to use rather than more difficult?) Now you can buy a router with a 4- or 5-port switch and wireless access point. You won't find many models that also include the cable or DSL modem built in along with these other services. That's because up until recently, there were differing standards for the broadband modems, and most of the ISPs used their own proprietary networking protocols. Today, most of the ISPs support the same standard broadband modems, so a universal network device for home use may already be available. Of course, when you are using a network switch backed by a router, you can also add new systems to the network by adding an additional switch in a hierarchical way. Most switches have an "uplink" port available.

Network Hubs

Network hubs were the earliest types of home Ethernet networking devices that made it possible to share files and play games in a LAN setting. Hubs are much slower than switches because a hub transmits each network packet to every other system in the LAN, whereas a switch establishes direct connections between each system in the LAN. Figure 5.3 shows a typical hub. Hubs are not as common today because switches are about the same price and provide faster network throughput.

Figure 5.3
An early 4-port 10-Mbps
network hub.

Figure 5.4 shows a hub from a different manufacturer that provides eight network ports. You could use this 8-way hub for a small workgroup in a work environment, or you could use it for a LAN party. Again, hubs suffer from poor performance compared to switches, which are comparably priced.

Network Switches

Network switches were once available only to businesses because they were so expensive. Companies like Cisco manufacture powerful network switches that are programmable at a low level (and Cisco switch programming is a black art known only to a few). In recent years, switches became available at the retail level. The earliest switches were fairly expensive: $150–200 for a 4-port switch. Now they are much more affordably priced, down at the sub-$50 range in some cases for a good 4-port switch. Figure 5.5 shows a typical switch from NETGEAR.

Figure 5.4
A newer 5-port 10/100 network hub.

Figure 5.5
Network switches provide active-dynamic connectivity.

Network Routers

You only need a network router if you have a broadband cable or DSL modem and you want to share that broadband connection with your entire network. Without the router, you do not have access to hardware NAT, which is required for online gaming. Figure 5.6 shows a typical router.

Figure 5.6
Network routers provide a gateway to a broadband device such as a cable modem.

Broadband: Cable and DSL

Most of the network equipment manufacturers have a wide range of products that serve both consumers (low-cost products) and corporate and government clients (for higher-end products). A typical broadband modem today (such as the model shown in Figure 5.7) supports both cable and Digital Subscriber Line (DSL) broadband connections.

Figure 5.7
D-Link is one of many cable/DSL modem manufacturers.

Figure 5.8 shows an all-in-one device: a cable modem with built-in router, 4-port 10/100 switch, and wireless access point. This is a great choice if you are building your first home network.

Figure 5.8
Linksys EtherFast cable/ DSL router with 4-port 10/100 switch.

Wireless Adapters

The standard Xbox 360 wireless adapter (shown in Figure 5.9) is a USB device that plugs into the back of your Xbox 360. This adapter is fairly expensive by wireless device standards today, with a retail price of $100 at the time of this writing. If you are tempted to buy a $30–40 USB wireless adapter instead, don't bother—it won't work without the proper drivers. Microsoft has provided the driver for its own wireless adapter, but there's no way to install a third-party driver on the Xbox 360. I tried several USB wireless adapters, from D-Link to Hawking to NETGEAR, and the Xbox 360 didn't recognize any of them.

Figure 5.9
The Xbox 360 wireless adapter supports 802.11a, b, and g.

However, there is a viable alternative to the expensive Xbox 360 wireless adapter. The trick is to tap into the network port directly rather than emulating a LAN connection through the USB port (which is how the wireless adapter works). You can use any device that provides a link through the Xbox 360's network port to a wireless network. There are many to choose from, and all are modestly priced (and they're half the price of the official wireless adapter).

One example that I am quite fond of is the Aeropad Mini wireless adapter, shown in Figure 5.10. I like this device because it utilizes an actual network cable tie-in, but it draws its power from a USB port. This device isn't using the USB port for a network connection because it connects to the LAN port on the back of the Xbox 360. However, it does use a USB port for its power source, so there's no need for a separate A/C power adapter. This is a nice option because this little Aeropad Mini can piggyback on your 360 like the official wireless adapter.

Figure 5.10
The Aeropad Mini wire-
less adapter draws power
directly from the USB port.

NETWORKING YOUR XBOX 360

If you are already experienced with PC networking, this will seem like familiar territory for you. Other console gamers may not have a background in networking, so I will go over the steps to hook up your Xbox 360 to your broadband connection—either directly to a cable modem or to a router or switch.

Beyond the Manual

If you are planning to share your broadband connection between your PC and Xbox 360, I strongly recommend using a network switch rather than a hub. Network hubs are fairly uncommon today, but it's an easy mistake to pick up a hub rather than a router. You want a router because Xbox Live games require a direct connection to the Xbox Live servers with NAT support. A switch has built-in NAT, but a hub does not, so you will most likely not be able to play games online if your Xbox 360 is plugged into a hub.

The Xbox 360 comes equipped with a built-in network port. The port is located on the back of your Xbox 360 on the right side (see Figure 5.11). A close-up view of the back of the 360 in Figure 5.12 shows the two I/O ports and the A/V port. The top I/O port is a USB port, whereas the bottom I/O port is a network port. Plug your network cable into this bottom port, and plug the other end into your router or switch.

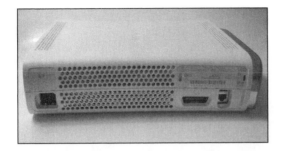

Figure 5.11
The rear of the Xbox 360.

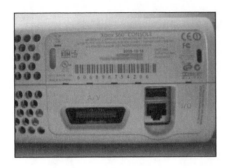

Figure 5.12
Close-up view of the network port on the back of the Xbox 360.

Configuring Your Xbox 360's Network Settings

Move the current folder selection in your Xbox 360's Dashboard to the far-right folder called System. This folder displays information about your Xbox 360 and allows you to change the configuration settings (see Figure 5.13).

After you select the Network Settings option in the System folder, you will see the screen shown in Figure 5.14. Move down to the Edit Settings option and select it. This brings up the Edit Settings screen shown in Figure 5.15.

You'll notice that the Edit Settings screen displays the current network configuration information. As you can see, my Xbox 360 is configured for Automatic IP settings. The IP here refers to the portion of the acronym TCP/IP, which stands for Transport Control Protocol/Internet Protocol. So, what we are looking at here are Internet Protocol addresses, or rather, IP addresses. The most common configuration for a home network is 192.168.0.x, where x represents each of the PCs and devices on your home network.

My 360 here has an IP address of 192.168.0.176. This value is completely meaningless on the Internet and is used only to identify my 360 within my home network. The subnet mask is another network configuration value that allows multiple domains to coexist when conflicting IP addresses might occur. The most common value for the subnet mask is 255.255.255.0. This is the same on every device on your network.

Figure 5.13
You can view and change
the network settings of
your Xbox 360 using the
System folder on the
Dashboard.

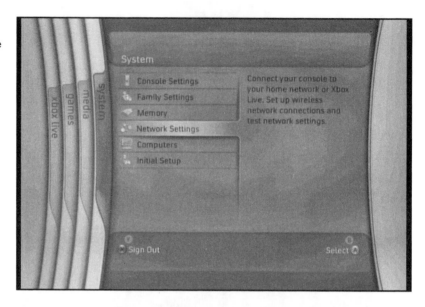

Figure 5.14
The Network Settings
screen.

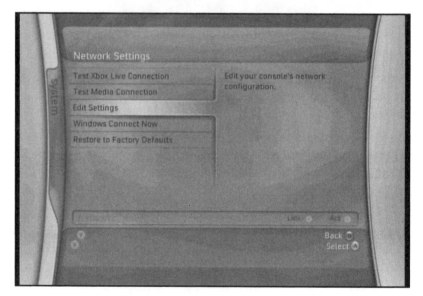

Next, look at the gateway value, which is 192.168.0.1. The gateway is the device on your network
that provides access outside the LAN—in other words, out to the Internet. In this instance, the
192.168.0.1 gateway address represents the router that is plugged into my cable modem.

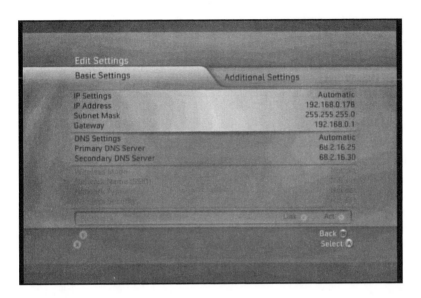

Figure 5.15
The Edit Settings screen displays the network configuration.

The second list of values deals with the Domain Name System (DNS) settings. Most broadband companies give you a primary and secondary DNS server. These server addresses are usually transmitted to your router through the cable or DSL modem.

If you need to set more advanced features manually for a custom network configuration (such as PPPoE), you can open the Additional Settings screen shown in Figure 5.16 to make changes as necessary.

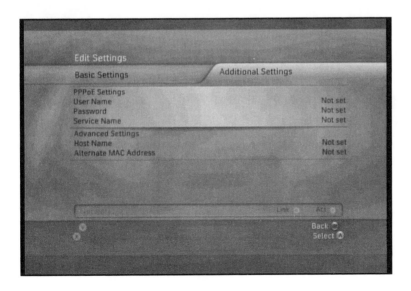

Figure 5.16
Additional network settings are configured here.

Manually Setting Your Network Configuration

If you select the IP Settings selection in the Network Settings screen, you can manually enter the network configuration values. People seldom do this today because network routers and switches provide this information to your Xbox 360 automatically when it powers up (when the network connection is established).

Figure 5.17 shows the screen that comes up when you choose to edit the IP settings. You have the option of setting it to Automatic or Manual at this stage. Automatic is what you should use most of the time; this tells the Xbox 360 to use Dynamic Host Configuration Protocol (DHCP) to automatically connect to the router or switch to configure the network. You need to set the values manually if you are using an old-style network hub instead of a router or switch. If you are connecting your Xbox 360 directly to the cable or DSL modem, the automatic (DHCP) setting should work; if not, you may need to refer to your broadband modem documentation for instructions on how to configure it.

Figure 5.17
Setting Automatic or Manual network configuration.

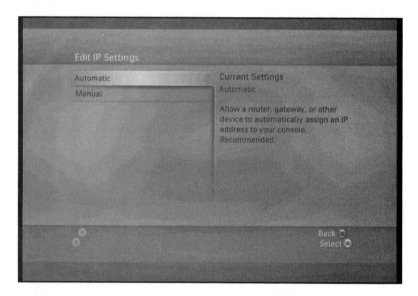

If you choose the Manual option, you see the screen shown in Figure 5.18. Selecting any option on this screen brings up the virtual keyboard that allows you to enter the new values. Note that the virtual keyboard shown in Figure 5.19 is limited to the numeric keypad because you need to enter only an IP address, not alphanumeric characters.

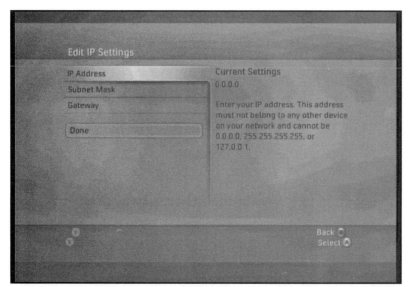

Figure 5.18
Configuring the IP settings manually.

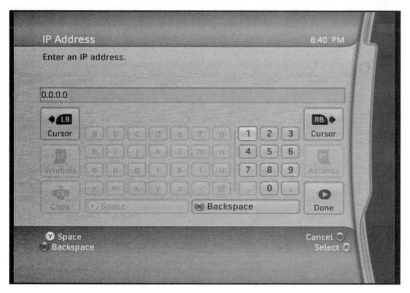

Figure 5.19
Entering the IP address manually.

Testing Your Network Connection

The Xbox 360 was designed to automatically configure itself for your home network, so connecting it should just be a matter of plugging in the network cable that came with the Xbox 360 and powering up your console. After you have networked your Xbox 360, you can do a lot of neat things by sharing files with the computers on your LAN. For instance, as you'll see in the next chapter, you can share music, photos, and videos stored on your PC with your Xbox 360; this is especially cool if you want to present these media files on your living room TV, perhaps with your surround system.

Let's check to see if your Xbox 360 has been properly configured for your home network. Open the Network Settings screen again (accessed from the System, Console Settings option). Move the selector down to Test Media Connection. This brings up the screen shown in Figure 5.20. Your Xbox 360 checks the network adapter (or wireless connection) and IP address and then tries to connect to a Windows PC on your network. This is a more advanced subject (sharing files with your PC), so it has been reserved for the next chapter.

Figure 5.20

Testing your Xbox 360's network connection.

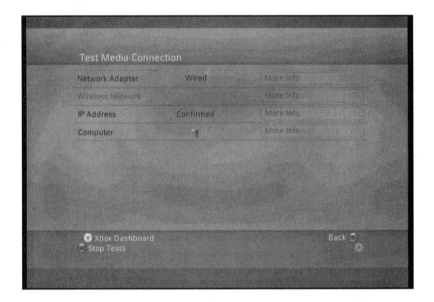

You can also go back to the Network Settings screen and try the first option, Test Xbox Live Connection for a more thorough test that verifies the Internet access from your Xbox 360 and then attempts to contact the Xbox Live server. This works even if you have not joined Xbox Live, because the test is only to determine if you can reach Xbox Live—you don't need to log in. While the test is taking place, you should see a screen that looks like Figure 5.21. When the tests have completed, you should see the screen shown in Figure 5.22.

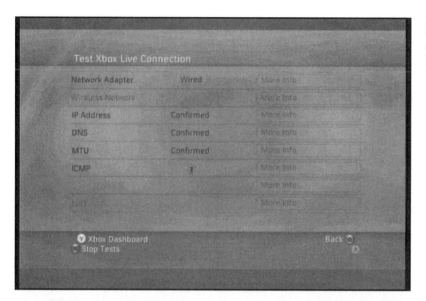

Figure 5.21
Attempting to contact the Xbox Live server.

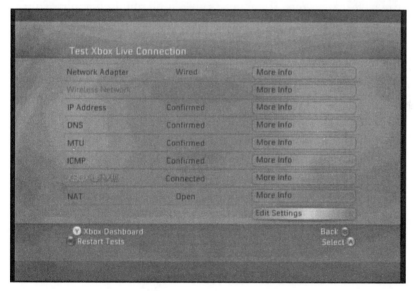

Figure 5.22
The Xbox Live server connection test succeeded.

HOSTING YOUR OWN LAN PARTIES

Suppose you're planning to host your own Xbox LAN parties and want to set it up so that your friends can bring their Xboxes over and simply plug into your network infrastructure without having to bring their own network gear (except for the requisite cable). If you have a broadband connection with a router already in place, you can share your Internet access with everyone at the LAN party—a big bonus for those who need to download new updates and patches for their games on Xbox Live. (Most games require that everyone is using the same version of the game.)

However, in the strictest sense, you don't need Internet access to host a LAN party. You only need to provide the networking infrastructure. (Broadband Internet often isn't needed.) I've hosted and attended many Xbox LAN parties, and I can tell you from experience that they are extraordinarily fun! Although it's technically possible to link four or more network switches (using their uplink ports), this often causes problems getting everyone connected because not every switch works well with every other switch. Your best option is to purchase a switch with a lot of ports built in. The most common high-end switches come in 8-, 16-, and 24-port models. They are relatively affordable (often priced under $100), so your home LAN party gear doesn't have to break your savings account.

Figure 5.23 shows a nice 24-port network switch made by Hawking and costing less than $100. There are similar switches available from D-Link, Linksys, SMC, NETGEAR, and others. If you don't need that many ports, you can opt for a 16-port or even 8-port switch to get the job done. Most original Xbox and Xbox 360 games support up to 16 players. But one benefit to having a 24-port switch is that you can run two LAN games at the same time, one with 16 players and another with eight. Or you can have three games of eight players each, all running simultaneously on your single 24-port switch.

Figure 5.23
This 24-port network switch should be able to handle your LAN parties on its own.

Supporting Wireless

If some of your friends have wireless adapters for their Xboxes, you may want to plug in a wireless access point into one of the free ports on your switch. However, for best performance, you should encourage everyone to use a wired connection to gain the full benefit of 100 megabits. Another problem with a wireless access point is that wireless players lose the benefit of a dedicated connection to other players—a key benefit when using a switch. I seriously doubt if a LAN game would run smoothly with most of the players connecting wirelessly, unless they are all in the same room and relatively close to the access point.

Network Cable Nightmare

One of the biggest problems at the LAN parties I've been to has been getting everyone hooked up to the switch from where they are set up in the house. If you're lucky enough to have a house big enough to handle a 32-player LAN (or have a friend with such a house), the biggest problem will be getting everyone wired into the switch. When you're dealing with such a large group of players, you probably won't be able to use a single switch because of the network cable nightmare that would ensue.

When dealing with a large number of players, you will probably be better off giving groups of players their own switch and then running a network cable from the uplink port on the switch to your main switch where the center of the action is in the house. If possible, keep it down to just one or two switches to improve performance, because all of those ports on a downstream switch will be sharing a single line instead of having their own lines.

"Switching" Connections

The real benefit of using a switch is that it establishes dedicated connections between every PC that is plugged in, providing for extremely fast network throughput. If you isolate some players behind a separate switch, they lose the benefit of that one-to-one connection with other ports in the switch. So, for the absolute best performance at a LAN party, try to get everyone in on the same switch if possible. In a heated LAN party battle with a game like *Halo 2*, that switch will be working insanely fast to send packets where they belong along all of those dedicated lines that it has established (and those lines are updated or terminated by the switch dynamically). For this reason, a switch can get extremely hot during a LAN party. Most rack-mounted switches in a large organization are stored in a climate-controlled room to keep them from overheating. If a router becomes too hot, you risk destroying the router. Heat is the primary cause of death for network and computer gear.

6

Sharing Media Files over the Network

If you intend to use your Xbox 360 merely for playing games, you are missing out on some fantastic features that this console is capable of providing for your entertainment. The Xbox 360 was not designed to be a Media Center PC, but rather as an accessory to one. Many new PCs are being shipped today with Windows XP Media Center Edition. This version of Windows is just another version of Windows XP, like the Home and Professional editions. Media Center Edition has special capabilities suited for your home entertainment center (including your TV and stereo surround system). These features include a custom "dashboard" suited for operation with a remote control that allows you to record broadcast, cable, and satellite TV; present slideshows of your digital photos in a scrapbook format; and play music files. Along with these features, Windows XP Media Center Edition can still run all your favorite applications and games, and it still has a Start menu. In this chapter, I will show you how to link your Xbox 360 to a Windows PC and explore some of the capabilities it has for sharing media files. You may find in time that you enjoy these capabilities as much as—if not more so—than playing games.

CONNECTING TO YOUR WINDOWS PC

If your Xbox 360 already has network access after you have configured it to connect to your Local Area Network (LAN), as described in the previous chapter, then your 360 can access shared media files on your Windows PC. To link up with your Windows PC from your Xbox 360, you need to install a small server program on your PC that knows how to communicate with your 360. The Xbox 360 is secured from normal network file sharing—to prevent hackers from manipulating the 360—so you can't just browse the LAN from your 360 for media files to play. Instead, you must configure the 360 to access a specific Windows PC on your LAN to share the files on that PC.

Windows XP Media Center Edition

Your Xbox 360 can play music files and view digital photos stored on a USB flash stick, digital camera, or CD-R/DVD-R disc containing media files. If you have Windows XP Media Center Edition on your PC, and it shares a network with your Xbox 360, you can do the following with it:

- Listen to your music collection
- View digital photos
- Watch live TV
- Pause and replay live TV
- Record TV shows for later viewing
- Watch videos
- Download and watch movies
- Buy music online

If you have just Windows XP Home or Professional, you can do the following:

- Listen to your music collection
- View digital photos

The list for Home and Professional is limited only in that these versions of the operating system have no support for streaming live TV and videos (which are delivered in the same manner).

The Media Center Edition link allows you to gain access to most of the features on your Media Center PC, such as news clips, downloadable music, and local weather. Specifically, your Xbox 360 can play videos from your Media Center PC with support for the following video formats:

- MPEG-1 with MPEG audio
- MPEG-2 with MPEG audio or AC-3 audio
- WMV 7, 8, and 9 with WMA Standard or WMA Pro audio
- WMV Image 1 and 2 (Photo Story 1, 2, and 3)

Your Xbox 360 can stream audio files from your Media Center PC from among the following audio formats:

- Windows Media Audio (WMA) Standard
- WMA Pro
- WMA Lossless
- MP3
- Any other format that provides a DirectShow decoder on the PC

Using the Media Remote

If you don't yet own a remote control for your 360—either the Premium or Universal remote—you may want to buy one for working with media files, because the Media Player was designed to work best from the remote control. Referring to Figure 6.1, there is a button on the remote control that looks like the Windows Start menu, with the word *Start* below it.

Pressing the Start button brings up the Media Player on your 360. (The launch screen is shown in Figure 6.2.) Visiting the link shown on this screen (http://www.microsoft.com/extender) brings up the Web site shown in Figure 6.3. This site gives you a wealth of information about Windows XP Media Center Edition.

Figure 6.1

The Universal Media Remote.

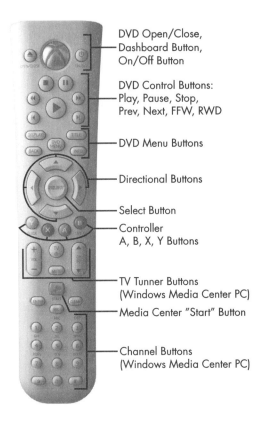

DVD Open/Close, Dashboard Button, On/Off Button

DVD Control Buttons: Play, Pause, Stop, Prev, Next, FFW, RWD

DVD Menu Buttons

Directional Buttons

Select Button

Controller A, B, X, Y Buttons

TV Tunner Buttons (Windows Media Center PC)

Media Center "Start" Button

Channel Buttons (Windows Media Center PC)

Figure 6.2

Pressing the Start button on the remote brings up this screen.

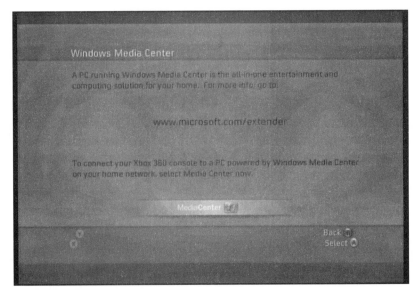

Figure 6.3
The Windows Media
Center 2005 Web site.

Launching the Media Player

Even if you don't have a media remote control, you can still open the Media Player from the
Xbox 360 Dashboard by navigating to the Media folder and selecting the Media Center option
shown in Figure 6.4.

Beyond the Manual

Microsoft Windows XP Media Center Edition 2005 is not available for retail purchase as an upgrade to your existing PC. You can only use this new operating system if you purchase a new PC
that comes with it. The Xbox 360 can still connect to Windows XP Home or Professional, but it
will lack the ability to stream live TV.

If you try the connection before you have configured your 360 to link to a Windows PC, you'll
see the screen shown in Figure 6.5. At the center of the screen is an eight-digit number that you

need to write down on a piece of paper because you must enter it when you configure your Windows PC as a host for your Xbox 360. This number uniquely identifies your Xbox 360.

Figure 6.4
Locating the Media Center in the Xbox 360 Dashboard.

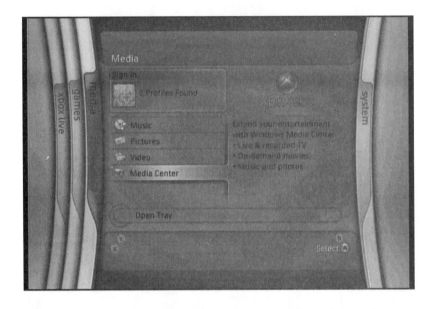

Figure 6.5
Pressing the Start button on the remote brings up this screen.

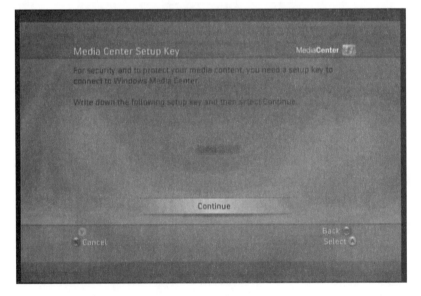

SETTING UP YOUR WINDOWS PC

You must download software on your Windows XP system to host media files that your Xbox 360 can access. This service is similar to a network file share, but it can stream media files to your 360. Streaming media hints of a Web page, and some parts of the media host program function more like a Web server than a file server. This server program takes up a small amount of memory on your Windows PC, but it is small enough to be unnoticeable and should have no impact on your system's performance unless your system is short on memory in the first place.

Beyond the Manual

I recommend 768MB or 1GB of memory for a Windows PC that will be sharing media files. These are common memory footprints for a system today that is often made up of one 256MB and one 512MB memory module, or two 512MB modules. Due to the continuously declining price of RAM modules and their increasing density, it is common to find average home PCs with 2GB of memory.

The next screen that comes up on the Xbox 360 (see Figure 6.6) instructs you to install the PC software on your Windows system (the PC that will host the media files to be streamed to your 360).

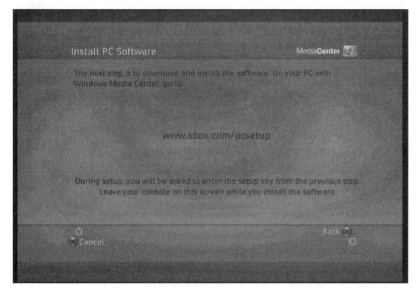

Figure 6.6
The Xbox 360 waits for you to prepare your Windows PC for media sharing.

Installing Windows Media Connect on Your PC

Next, go to your Windows PC and browse the Web to http://www.xbox.com/pcsetup to bring
up the Web site shown in Figure 6.7.

Figure 6.7
The Xbox 360 Media
Setup Web site.

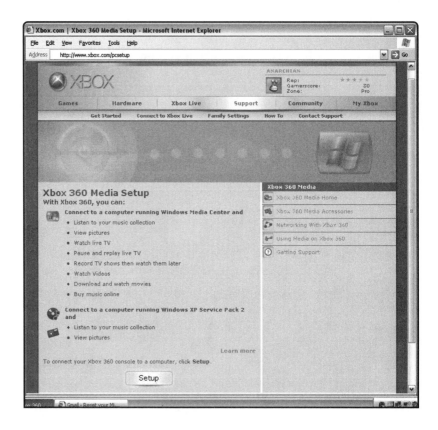

Click the Setup link at the bottom of this page to bring up the page shown in Figure 6.8, which
verifies that your PC is running Windows XP (Service Pack 2 or later) or Media Center Edition
2005 (or later). This page tells me that I can only stream music and pictures to my 360 because
I am not running Media Center Edition.

Clicking the Next link on this page causes the File Download dialog box to appear (see
Figure 6.9). You can choose to run the installer directly after download, or you can save the
file to your hard drive first and then run it manually. Note that the file name shown in this figure
will change depending on the version of Windows you are using.

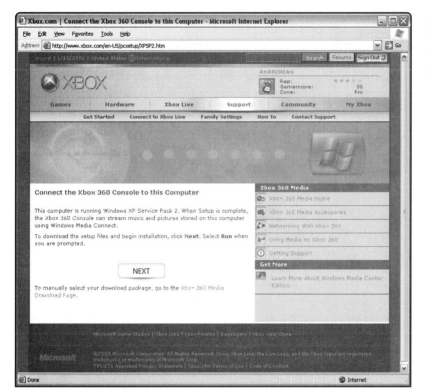

Figure 6.8
Verifying the version of Windows installed on your PC.

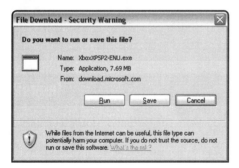

Figure 6.9
Downloading the Xbox 360 Media Connect software.

When the Xbox 360 Media Connect installer program runs, the screen shown in Figure 6.10 comes up on your Windows PC. This screen indicates that you should install this program if you are using Windows XP Home or Professional. Before continuing with the setup, make sure your Xbox 360 is turned on and is connected to the network. Then you can click the Next button. Note that this screen looks a lot like the Xbox 360 Dashboard, but this is actually a screen shot from the installer running on a PC, so don't let the similar appearance confuse you.

Figure 6.10
The Xbox 360 Media
Connect setup program
running on a Windows
PC.

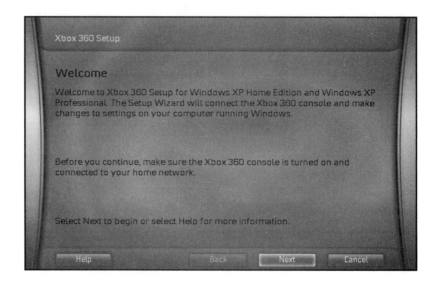

The next screen that comes up (see Figure 6.11) displays the license agreement that you must accept before the installer will continue. This End User License Agreement (EULA) specifies that you may only use it on a PC running a valid license of Windows. Select the I Accept option and then click Next.

Figure 6.11
The End User License
Agreement.

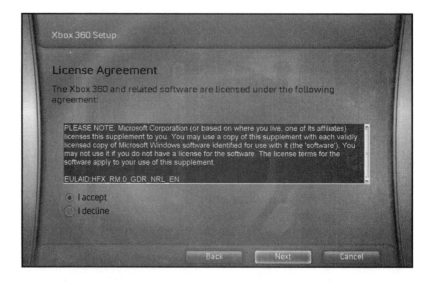

Next, the screen shown in Figure 6.12 comes up to notify you that the Windows Media Connect software is being installed on your PC.

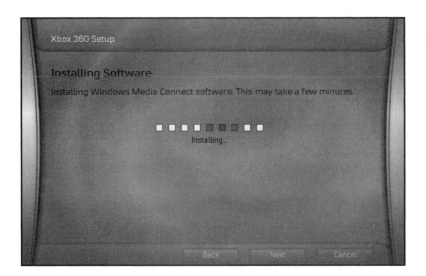

Figure 6.12
Windows Media Connect software is being installed.

Configuring Windows Media Connect

After the Windows Media Connect software has been installed on your Windows system, it's time to configure it. The next screen that appears (see Figure 6.13) notifies you that system standby and hibernation modes prevent your Xbox 360 from connecting to the PC. Most Windows systems are configured so that they automatically go into suspend mode whenever the system has been left alone for a certain period of time—usually one hour. This not only conserves electricity but also extends the life of your PC.

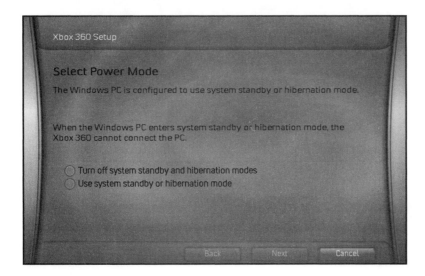

Figure 6.13
Selecting the power mode options on your Windows PC.

This is a power-saving feature meant to conserve electricity when your PC is not in use. You configure this feature through the Screen Saver tab in the Display Properties program (via the Control Panel). Power-saving features are considered part of the screen saver functionality in Windows. Hibernation support means that after a longer period of time has elapsed with no activity on a Windows PC, the computer will store all memory to a temporary file on the hard drive and then power down.

When the system is powered back up again, the memory is restored using the temporary file and the system resumes as if it had never been turned off. Obviously, these features disrupt any file sharing on the system, and that includes hosted media files accessed by your Xbox 360. If you will frequently be using your Xbox 360 to stream files from your PC, I recommend turning off system standby and hibernation modes. But if you only occasionally share media files from your PC, you can go ahead and use system standby and hibernation modes.

Testing the Connection to Your Xbox 360

The next screen that comes up (shown in Figure 6.14) is a progress screen that tells you the Xbox 360 is being contacted for authorization to share media with your Windows PC. When completed, the program tells you that it has located and authorized your Xbox 360 (see Figure 6.15).

Figure 6.14
Authorizing the Xbox 360 with this Windows system.

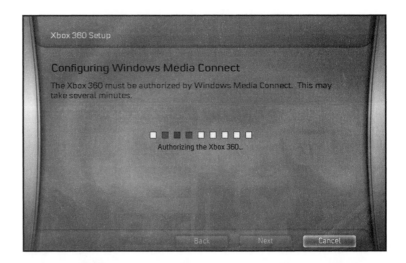

Geek Speak

Authorization in this case simply means that your Windows PC is trying to locate your Xbox 360 on the network to set up a communication link between the two.

Figure 6.15
Windows Media Connect software has located the Xbox 360 on the network.

Clicking Next brings up the screen shown in Figure 6.16, notifying you that the Xbox 360 setup has completed.

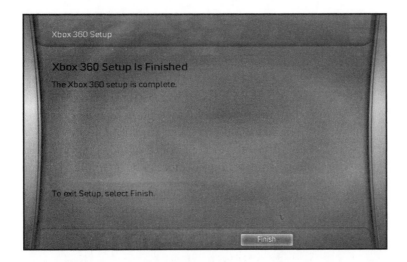

Figure 6.16
Windows Media Connect has now been set up to communicate with your Xbox 360.

Beyond the Manual

You can change the Windows Media Connect settings using the Control Panel on your Windows PC.

At this point, the Xbox 360 is still waiting for Windows Media Connect software (running on your Windows PC) to notify it that media sharing is now available (see Figure 6.17). Because I am not using Windows XP Media Center Edition, this screen will never change on my Xbox 360. I need to go to the Media folder in the Dashboard and access music and pictures through that interface instead. If you have Windows XP Media Center Edition on your PC, you will be given access to the Media Player features on your 360.

Figure 6.17
The Xbox 360 is still waiting for a live connection to Windows Media Connect.

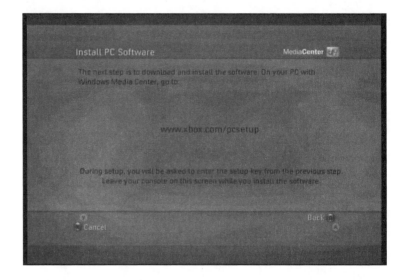

Testing Windows Media Connect

How will you know if Windows Media Connect is actually running? If you look on your Windows system at the right side of the taskbar (which shows small icons of services running on your system, such as anti-virus software), you will see a new icon. See Figure 6.18.

Figure 6.18
The Windows Media Connect service is running on my Windows PC.

You can double-click the Windows Media Connect icon in the taskbar to bring up its configuration program. Figure 6.19 displays the main tab of this configuration program showing the devices that are connected to the PC.

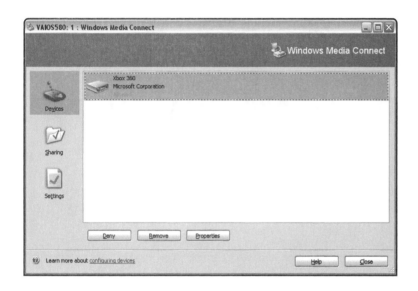

Figure 6.19
The Windows Media Connect configuration program showing connected devices.

Alternatively, you can bring up the Windows Media Connect configuration program through the Windows Control Panel, which is accessed from the Start menu. The Control Panel is shown in Figure 6.20, with Windows Media Connect highlighted.

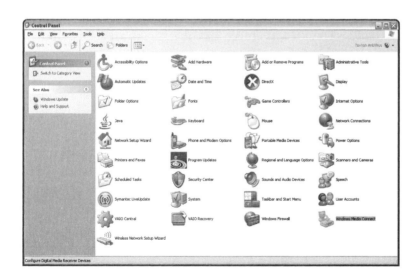

Figure 6.20
Bringing up the Windows Media Connect configuration program using the Control Panel.

SHARING FILES WITH YOUR XBOX 360

When you've completed the preceding steps, you can proceed to access your Windows XP Home, Professional, or Media Center Edition PC from your Xbox 360 to stream media files.

Sharing Music Files with Your Xbox 360

Let's browse the remote Windows PC from our Xbox 360 to stream some music files. Open the Media folder in the Dashboard and select the Music option, shown in Figure 6.21.

Figure 6.21
Preparing to browse for music files over the LAN.

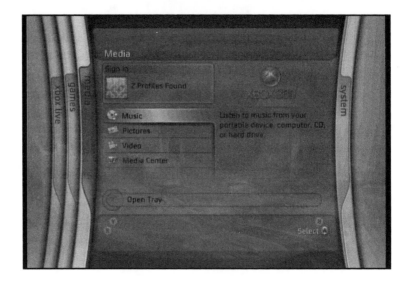

The Music screen, shown in Figure 6.22, shows the list of available sources for music files. Because I have not inserted a CD-R or DVD-R with music files on it, the Current Disc option is disabled. Likewise, Music Player is disabled because I haven't started playing music yet. After I've started playing music (from any source), this option becomes available. The Portable Device option is also disabled because I have not inserted a USB storage device.

When you select Computer as the source for music files for the first time, you will see the screen shown in Figure 6.23. This is a verification asking whether you have previously set up your Windows system with the Windows Media Connect software. Because you just completed that step a few minutes ago, go ahead and select Yes, Continue.

When you select this option, the Xbox 360 first goes out on the LAN (see Figure 6.24) to locate the Windows PC you have configured for use with Windows Media Connect software.

When the Xbox 360 locates an instance of Windows Media Connect software running on a Windows PC on your LAN, it displays the screen shown in Figure 6.25. The name of my PC in

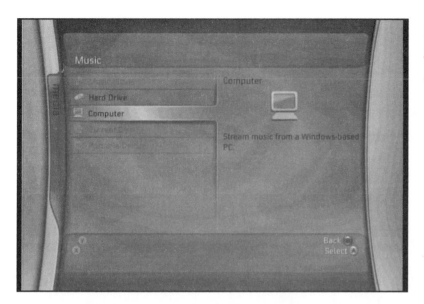

Figure 6.22
Browsing a remote Windows XP system on the LAN.

Figure 6.23
Preparing to access the remote Windows PC.

this example is VAIOS580. Note that this is the actual name of the PC that is broadcast to other computers on the network.

Figure 6.24
Xbox 360 is searching the LAN for your Windows PC.

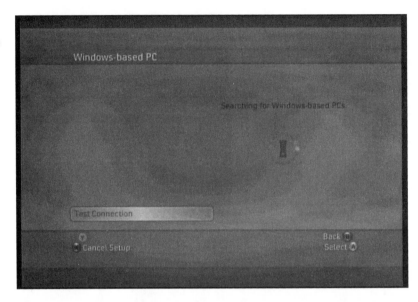

Figure 6.25
An instance of Windows Media Connect has been located on the networked PC.

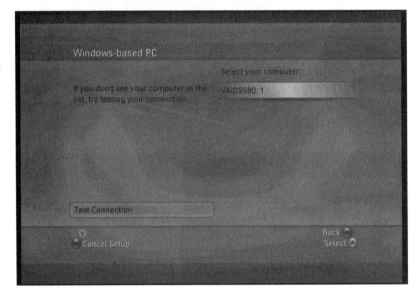

First, you must select the Windows PC that you want to connect to. In this example, I have selected VAIOS580. The screen shown in Figure 6.26 appears. The Xbox 360 has downloaded a listing of music files from Windows Media Connect, sorted by album, artist, and so on.

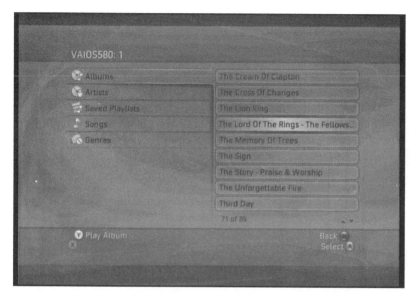

Figure 6.26
The Xbox 360 is browsing music on the networked PC.

Selecting an album brings up the screen shown in Figure 6.27, with the songs in this album displayed on the list at the right. This is the soundtrack from *The Lord of the Rings: The Fellowship of the Ring* by Howard Shore.

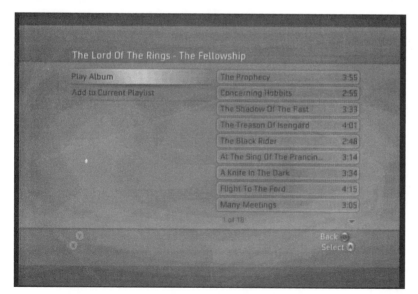

Figure 6.27
Preparing to play an entire album of music.

When you select Play Album, the entire album is added to your current playlist and playback begins (see Figure 6.28). Alternatively, you can add selected files from the album to your playlist. You can also continue browsing music files and selectively add them to the current playlist—even while music is currently playing.

Figure 6.28
Playback of the album has begun; a graphic visualization follows the tempo of the music.

After establishing this link with your Windows PC on the network, you can browse the music files on that remote PC at any time, and it will be treated abstractly by the 360 as if it were just another type of storage device (such as a USB flash stick).

Sharing Photos with Your Xbox 360

You can also view digital photo slideshows that are streamed from your Windows PC to your Xbox 360. From the Media folder in the Dashboard (shown in Figure 6.29), select Pictures.

The Pictures screen comes up, as you can see in Figure 6.30. Select the Windows PC that is shown, or you may need to establish a connection again.

When you select the remote Windows PC, your Xbox 360 communicates with the Windows Media Connect server software running on that PC, which transmits the directory structure within the My Pictures folder on the remote PC. So, if you want to share pictures with your 360, put them in My Documents, My Pictures (see Figure 6.31).

Select a folder or any digital photos located in the main My Pictures folder (see Figure 6.32) to view a slideshow of those photos (see Figure 6.33).

132

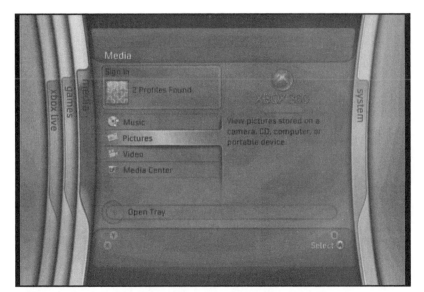

Figure 6.29
The Media folder in the Xbox 360's Dashboard.

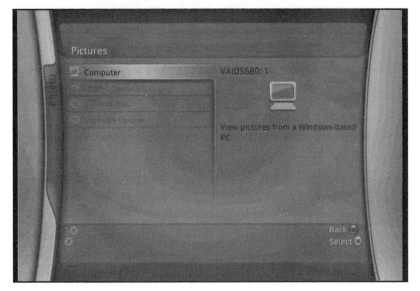

Figure 6.30
Preparing to browse some digital photos on the remote Windows PC.

Figure 6.31
The list of picture folders
found on the remote
Windows PC.

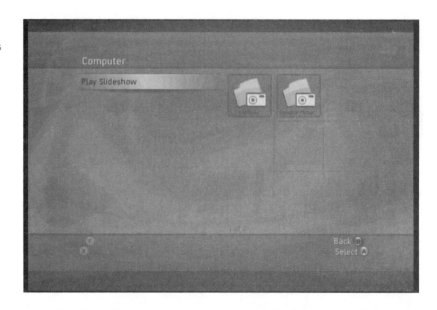

Figure 6.32
Getting ready to view the
slideshow.

Figure 6.33
The photos are sent to
your Xbox 360 in
slideshow format.

the
gadget
geek's
guide to
Your Xbox 360

7

Going Online with
Xbox Live

In this chapter, I will assume that you have never signed onto Xbox Live before. I will take you step by step through the process of creating an Xbox Live account so that you can begin playing games online.

CREATING YOUR XBOX LIVE ACCOUNT

First you need to make sure that you are connected online. You should have taken care of this back in Chapter 5, "Connecting Your Xbox 360 to Your Home Network." If you skipped that chapter and are not wired yet, go back and read that chapter to get your 360 connected.

There are only a few differences between using a hard drive and using a memory unit, so I will not go into those differences here. Just note that some of the features described in this chapter may not be available without a hard drive—although you will still be able to get the full functionality of Xbox Live using a memory unit. The sections of this chapter (and most of the remaining chapters in the book) assume that you have a hard drive because I discuss the downloading of large files such as movie trailers and game demos, in addition to updates for your current games.

Signing Up

Let's start by going to the Xbox Live folder in the Dashboard (the leftmost screen). Move the selector down to the Xbox Live box, as shown in Figure 7.1.

This selection brings up the Xbox Live sign-up screen shown in Figure 7.2. You need to go through this screen even if you are already an Xbox Live member because of your original Xbox account.

Figure 7.1
The Xbox Live folder is part of the Dashboard.

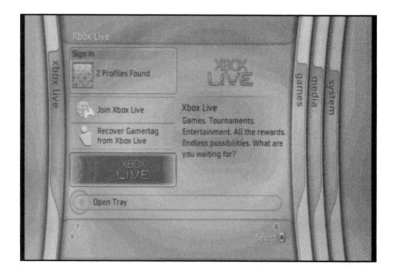

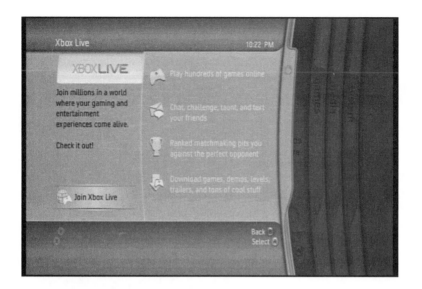

Figure 7.2
The Xbox Live sign-up
screen.

Updating Xbox Live

Because I'm walking you through the process step by step, you should see the same screens I'm showing you here on your TV, although some of these features are likely to change over the next few years. Xbox Live is frequently updated to incorporate new features and new games, so the first thing you must do before signing up is download the latest version of Xbox Live. When you see the screen in Figure 7.3, select Yes, Update Now.

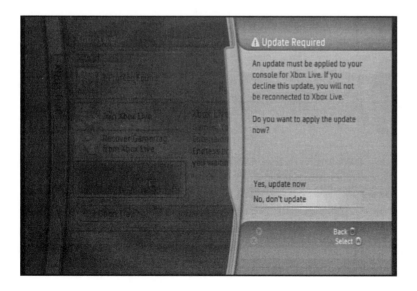

Figure 7.3
You must update Xbox Live before you can sign up.

The update for Xbox Live downloads (see Figure 7.4) and installs automatically, which allows you to sign up.

Figure 7.4
The update to Xbox Live is downloading.

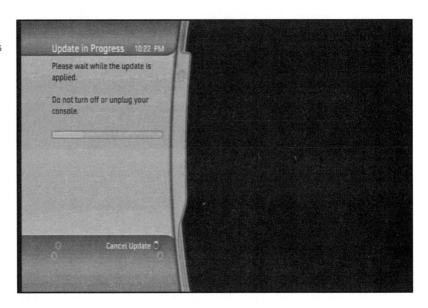

Connecting to Live

The actual connection process is automatic, because you only have to provide a password the first time you sign on to Xbox Live. Your account information is retained on the memory unit or hard drive in your Live account file. First, you see the screen shown in Figure 7.5, which is displayed while your Xbox 360 attempts to reach the Live server. If there is a problem with the connection at this early stage, it is most likely due to your network configuration. (Refer to Chapter 5, "Connecting Your Xbox 360 to Your Home Network.")

Assuming that your Xbox 360 is plugged into your network correctly (unless you have plugged it directly into your broadband adapter—either a cable modem or DSL modem), the next step that your 360 performs is to check for updates to Xbox Live, as shown in Figure 7.6. This process actually took place just a few moments ago when you first tried to sign up and the update was downloaded. Because your Xbox 360 now has the latest version of Xbox Live, this screen flashes by quickly.

The next thing that happens—if this is the first time you have signed on to Xbox Live—is that your offline Gamertag is sent to Xbox Live for verification, as shown in Figure 7.7. More than likely, unless you typed in an unusual Gamertag when you first powered up your Xbox 360 (back in Chapter 1, "Firing Up Your Xbox 360"), Xbox Live will require you to select an alternate

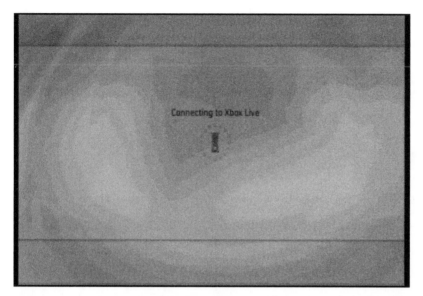

Figure 7.5
Connecting to the Xbox
Live server.

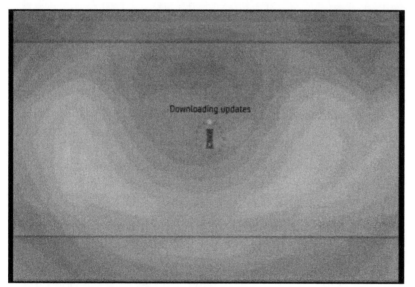

Figure 7.6
Checking for updates to
Xbox Live.

nickname. Because hundreds of thousands—perhaps even millions—of players have signed up already, all of the most common nicknames have been chosen already.

In this case, the Gamertag I've been using while playing games offline—ENIGMA—has been taken already. This is not surprising because that is a fairly common nickname (or *handle*) online.

Figure 7.7

Logging in using your Gamertag.

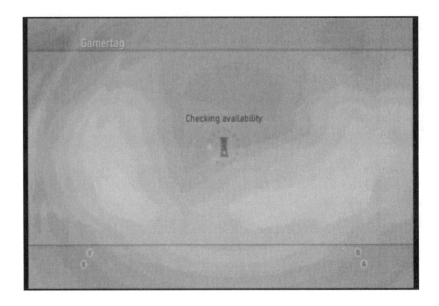

You are not likely to have luck with your favorite character names either, because they will have all been chosen already. Keep in mind that Xbox Live was in operation for four years at the launch of the Xbox 360. Gamertags like Neo, DarthVader, Pokémon, and so on will have already been chosen. Of course, if you use an instant messenger service like AIM or MSN Messenger, you are already familiar with these nickname limitations and will be expecting to see Figure 7.8.

Geek Speak

A *handle* is a nickname used to represent yourself when interacting with others online.

You can ask Live to recommend some alternate Gamertags for you, or you can create a new one. You will be asked to enter a new Gamertag using the virtual keyboard on the screen. This is where a USB keyboard saves you a lot of time, because this is only the first of many text fields you have to enter while signing up to Xbox Live. You might just plug your PC keyboard into the Xbox 360 temporarily while entering data; if it's a standard USB keyboard, the 360 should recognize it without complaint. I recommend a wireless keyboard with a USB transceiver so that you can pull out the keyboard any time you need to use it. If you are already using the two front USB ports for your controllers, you can use the extra USB port on the back of the 360 that is unused unless you have the wireless network adapter.

In Figure 7.9, you can see that I've chosen an unused Gamertag that Xbox Live accepted, and I am now asked if I have a Passport Network account. What is a Passport Network account?

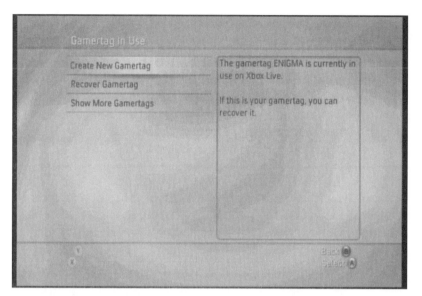

Figure 7.8
My offline Gamertag is already in use, so I'll need to enter a new one.

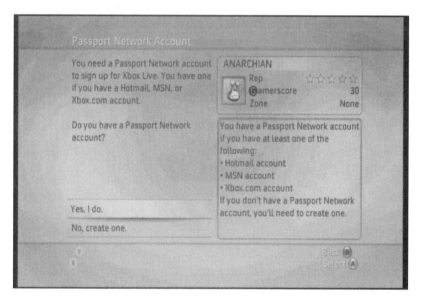

Figure 7.9
Do you have a Passport Network account?

The Passport Network is something that Microsoft has been pushing for several years now. No one seems willing to adopt it, so Microsoft has tried to utilize Passport within its own products—for product registration, e-mail accounts, and so on.

Beyond the Manual

When you create an Xbox Live account, your offline Gamertag information is migrated to your Live account with your Gamerscore. Your gaming skill is tracked using your Gamerscore, and 360 games are designed to add to your Gamerscore when you accomplish things like completing a level or unlocking a new car or weapon (depending on the game). Your online and offline score is one and the same. In a game like *Project Gotham Racing 3*, you will be seriously disadvantaged online if you haven't unlocked the best cars in each performance class!

Passport was conceived during the development of Microsoft .NET, which is a development platform for writing software for Windows and the Web. The .NET Framework is a giant hierarchical library for software development that provides—presumptuously—everything you could possibly need to write software for Windows or the Web.

The biggest problem with Passport is its centralized nature. Who wants to use Microsoft servers for their own confidential information? Passport was something of a backpack full of bricks for many years that users of Hotmail and MSN Messenger have had to put up with to use these services—which are, arguably, free services, so one cannot exactly complain. Now that Passport is being pushed through Xbox Live, there is a real possibility that Microsoft is merging the PC and video game console into one seamless world.

The result may very well be that you will be able to buy, sell, and trade game resources online with other Live/Passport users, and this information will also be available on the Web. What might be some of the repercussions of this technology? How about a partnership with eBay.com, bringing a customized interface to Live that deals only with auctions specifically related to Xbox 360, with the ability to use Microsoft Points to pay for auctions (using a sort of automatic monetary exchange with PayPal)? Now *that* would be interesting!

In any event, let's go ahead and create a new Passport Network account. This entails several steps, each requiring you to enter data in a number of screens, which is why I recommend a USB keyboard for your convenience.

First, you are asked for your locale, as shown in Figure 7.10. Then you are asked to enter your birth date as a means of account verification later on (see Figure 7.11). Note that if you are under the age of 18, you must have a parent or guardian set up a Passport account for you.

The next screen that comes up, shown in Figure 7.12, asks for your e-mail address. This e-mail address will be used to send you information about your Xbox Live account, so be sure to enter it correctly.

The next screen that comes up, shown in Figure 7.13, asks you to enter a password (with verification) and a secret question with the answer to be used in case you forget your password.

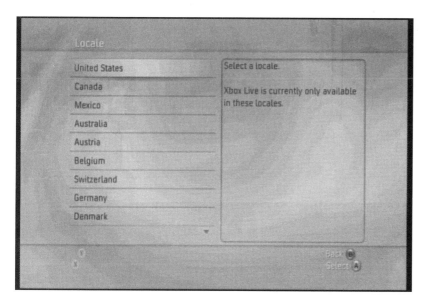

Figure 7.10
Selecting your geographic location.

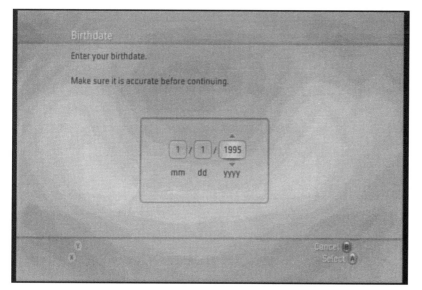

Figure 7.11
Entering your birth date.

Figure 7.14 shows the virtual keyboard that you use to type information into text fields. You can type directly into a text field using a USB keyboard, which is especially helpful when entering hidden password characters.

Figure 7.12

Enter your e-mail address.

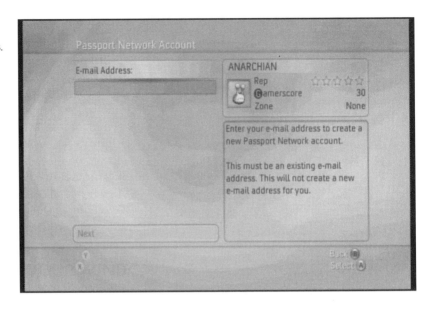

Figure 7.13

Enter your password with verification information.

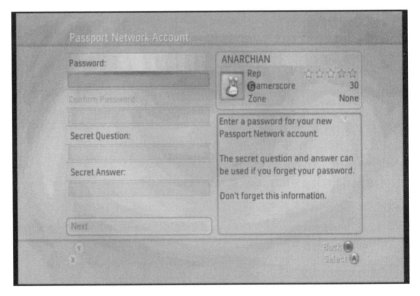

Now that you have completed some of the logistical information used to sign up and so forth, it's time to enter personal information. Figure 7.15 shows the screen you use to enter your name and phone number. This information is kept confidential and will never be shared with other gamers on Live. If you feel uncomfortable sharing this information, you can enter a bogus phone

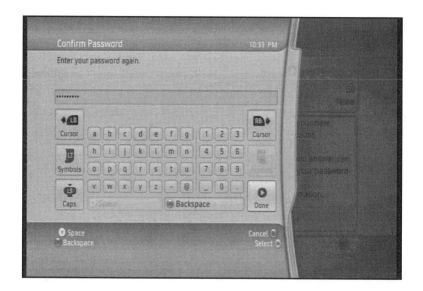

Figure 7.14
The virtual keyboard.

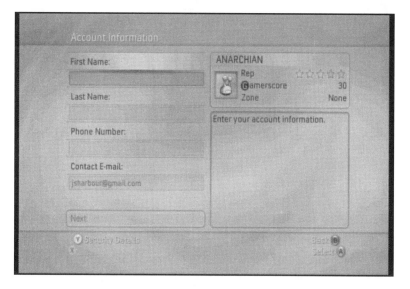

Figure 7.15
Enter your name and phone number.

number, but your real name will be required for billing (unless you use a prepaid gamer card, which you can purchase at retail video game stores).

After you enter all the necessary information in the preceding screens, you are presented with the Xbox Live Terms of Use and Privacy Statement (shown in Figure 7.16). When you accept the Terms of Use agreement, you see the screen shown in Figure 7.17.

Meltdown

Among other things, the Terms of Use agreement states that if you cheat or use a modified Xbox, your Xbox Live account will be terminated without the benefit of a refund. This was a big issue with the original Xbox.

Figure 7.16
Accept the Terms of Use and Privacy Statement to continue.

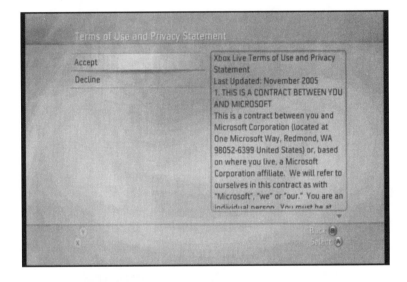

Figure 7.17
Your Xbox Live/Passport Network account has been created.

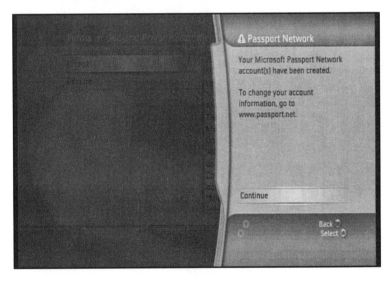

Are You an Electronics Geek?

In 2002, shortly after the release of the Xbox, an MIT student by the name of Andrew Huang applied his electronics engineering knowledge to decode the encryption algorithm used to secure the system bus of the Xbox. Shortly after he published his work, the first Xbox mod chips became available (by another party). Andrew Huang is a hardware hacker, which does not mean he is a criminal (although the incompetence of the media would have you believe otherwise). He wrote about the experience in his book *Hacking the Xbox: An Introduction to Reverse Engineering* (ISBN 1593270291), published in 2003 through a legal shield called Xenatera LLC. This book explains the architecture of the Xbox in fine detail, which likely brought Mr. Huang under the crosshairs of the Microsoft legal guns. There is a law in the United States called the Digital Millennium Copyright Act (DMCA) that is frequently used to shut down people like Mr. Huang.

How deep does the rabbit hole really go? Andrew Huang is a close friend of Joe Grand (http://www.grandideastudio.com) who published a book in 2004 titled *Game Console Hacking: Have Fun While Voiding Your Warranty* (ISBN 1931836310). Joe is a fellow instructor (part time) at the University of Advancing Technology (UAT), the campus I now call home (http://www.uat.edu). Although I've never met Andrew Huang, I feel a kinship with him through Joe Grand and through the book I wrote in 2004 titled *The Black Art of Xbox Mods* (ISBN 0672326833). This book does not explain the Xbox architecture; it is a guide for non-engineers on how to modify an Xbox—essentially to turn it into a $150 media center for your living room. Although Microsoft is concerned about what people are doing with the Xbox, only a small percentage of consumers are modifying their Xboxes, and these changes render the Xbox unusable on Live.

Selecting a Payment Method

The next step to the Xbox Live sign-up process is to choose a payment method or membership type. Figure 7.18 shows that several options are available:

- Monthly Gold Membership
- 3-Month Gold Membership
- 12-Month Gold Membership

The one option you do not see in this figure is at the bottom of the list. (Note the small arrow at the bottom of the list.) By moving down the list, you reveal another option that is not obvious to the casual user's eyes. The hidden option, Xbox Live Silver Membership, is shown in Figure 7.19.

Figure 7.18
Xbox Live membership
subscription options.

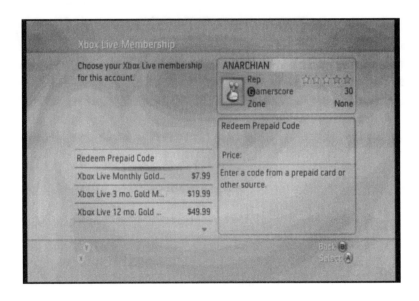

Figure 7.19
Locating the no-cost Silver
Membership option.

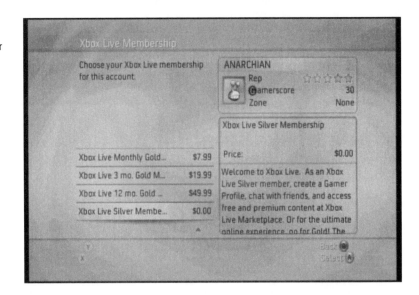

If you already have an Xbox Live Gold Membership (perhaps from your original Xbox), the Silver Membership option is not available, or at least, it won't be unless your Gold Membership expires. This type of membership allows consumers to browse the offerings on Live, such as the Marketplace (see Chapter 8, "Xbox Live Marketplace: Game Demos, Trailers, Themes, and Microsoft Points") and Live Arcade (see Chapter 9, "Xbox Live Arcade: Downloading "Casual" Games")

without requiring a subscription. Any game that has online multiplayer gameplay requires a Gold membership, including downloadable games.

It's inevitable that you will need a Gold membership, but the Silver option at least allows you to get a feel for Xbox Live before you commit to it. Fortunately, new subscribers have a free 30-day trial of Gold available, but this, too, is not very obvious. When you select the Silver Membership option, you are presented with the screen shown in Figure 7.20. This screen gives you one more opportunity to sign up for a paid subscription model before revealing the free 30-day trial Gold membership. Go ahead and select Keep Silver at this point.

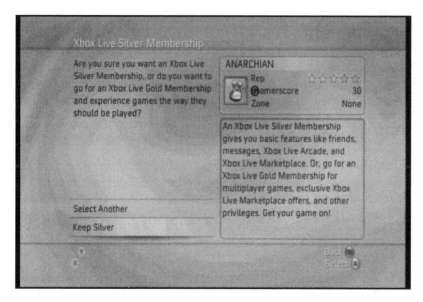

Figure 7.20
The Microsoft share-holders are counting on you.

You then have the opportunity to accept the free one-month trial of the Gold Membership that is offered to new subscribers, as shown in Figure 7.21.

Beyond the Manual

If you are already a member of Xbox Live via your original Xbox, you can continue using that account with your Xbox 360.

Figure 7.21
Taking advantage of the free one-month trial Gold membership.

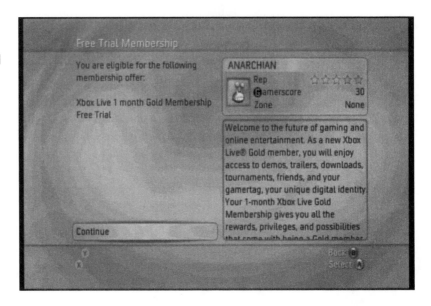

Entering Your Address

Next you are asked to enter your address into the screen shown in Figure 7.22. This information is required for billing purposes. If you don't want to be billed directly by Microsoft, you can purchase a prepaid Xbox Live Gold card at a retail store and apply the time credit on the card to your account.

Choosing Your Gamer Zone

The next screen is an interesting one because it affects the types of players that Xbox Live pits against you when you play online. As you can see in Figure 7.23, there are four zones to choose from:

- Recreation
- Pro
- Family
- Underground

You must follow certain rules in each of the four zones. Foul language is not permitted on the Recreation or Family zones. Other players could report you, which would affect your reputation negatively. The Pro zone is not as restrictive as long as you treat other players respectfully. The Underground zone is a no-holds-barred area where anything—including bad behavior—is tolerated.

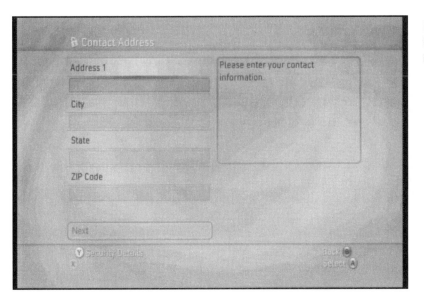

Figure 7.22
Entering your address information.

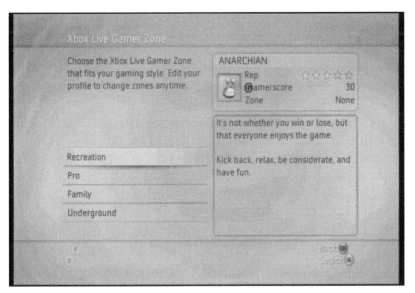

Figure 7.23
Choose the game zone that best suits your play style.

Recreational Gaming

The Recreation zone on Xbox Live is for casual gamers who are not concerned with high-score lists, leaderboards, and that sort of thing. If you just want to enjoy your online gaming experience with friends—while making new friends—this is the zone for you.

Pro Gaming

The Pro zone on Xbox Live is for professional gamers who know what they're doing and don't want to deal with newbies (beginners). If you are new to Xbox Live but have much experience playing games online using your PC, you may feel more at home in the Pro zone than in any of the others. Then again, you might get humbled and need to back out to the Recreational zone to protect your self esteem.

Family Gaming

The Family zone should be the safest zone for kids to use online. Although there is no voice filter technology to protect young ears from the threat of foul language, the reputation factor is of serious concern to all gamers, and you have the option to add any gamer to your ignore list. The key here is to keep an eye on your children when they are playing online. You may want to turn off voice chat altogether to avoid the issue if you are not able to supervise your child while he is playing online.

Underground Gaming

The Underground zone is "not for the faint of heart" according to the Xbox Live description. This zone is where gamers can go if they become emotionally charged during an intense game and want to let loose without worrying about offending anyone else. Like it or not, you have a responsibility to be courteous to others, which means staying out of the other zones if you aren't able to play nicely and hold your tongue.

Xbox Marketing

The next screen that comes up is Xbox Marketing. It asks whether you'd like to allow Microsoft to spam your e-mail account with junk mail (see Figure 7.24). If you are truly interested in receiving advertisements related to video games, go ahead and accept the offer; otherwise, you may decline. Actually, I am being only half serious. Most of the content sent to you will be directly from Xbox Live notifying you of new updates.

Completing the Account Creation Process

The next step is for Xbox Live to save all the information you've entered and create your new Live account (see Figure 7.25). After you've saved the account information and created your account, you see the Congratulations screen shown in Figure 7.26.

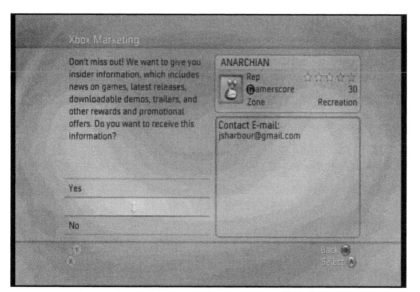

Figure 7.24
Would you like to receive nicely packaged spam e-mail?

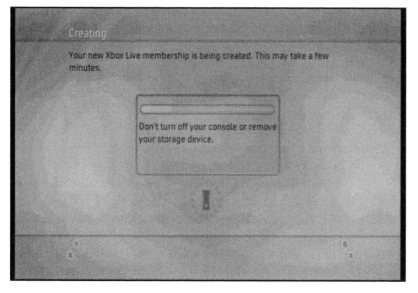

Figure 7.25
Your new Xbox Live account is being created based on the information you've entered.

Figure 7.26
Your new Xbox Live account has been created.

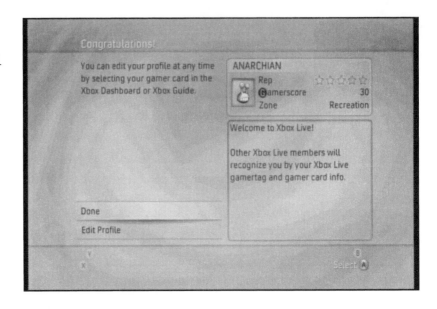

As soon as you select Done, you are taken back to the Xbox Live folder in the Dashboard, shown in Figure 7.27.

Figure 7.27
The Xbox Live folder in the Dashboard.

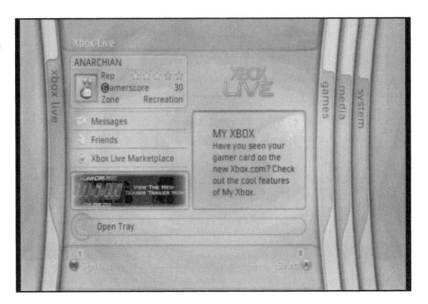

TESTING XBOX LIVE

Let's take Xbox Live for a quick spin to see how fast your connection is and to get a glimpse of some of the things you'll find online (which are covered in the next few chapters). Figure 7.27 showed you the Xbox Live folder of the Dashboard. This folder frequently displays small game-related advertising windows. Check out Figure 7.27 again, and look at the black box just above the Open Tray entry. This is an advertisement for an upcoming movie, *Mission Impossible 3*. Move the pointer to this image and select it. (Note that the link you select will probably be different from what I'm showing you here.)

Downloading a Movie Trailer Video

Figure 7.28 displays the screen that comes up. This screen shows details about the movie trailer, revealing that it is 85.73MB in size and runs at 720p. (There are 480p versions of most videos on Live.) Notice on the left-hand side where the price is shown that this is a free download. I have no Microsoft Points. Because this trailer is free, you can move down to the option Confirm Download.

Beyond the Manual

You'll learn about Microsoft Points and purchasing online content in the next two chapters.

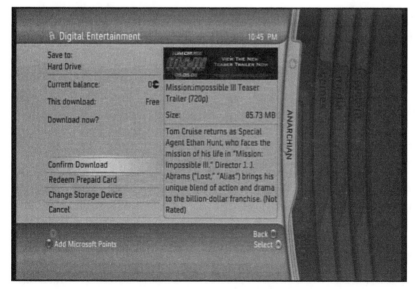

Figure 7.28
This Digital Entertainment download screen shows details about the content item.

The movie trailer starts downloading, as you can see in Figure 7.29. I think it's rather obvious that you won't be able to download content of this type unless your Xbox 360 is equipped with the hard drive accessory. This video clip alone demonstrates why, because it is far larger than an entire memory unit (which provides only 64MB of storage).

Figure 7.29
Downloading a movie trailer from Xbox Live.

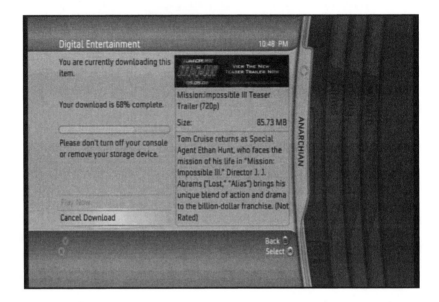

Watching the Movie Trailer Video

When the download has completed, you are presented with the screen shown in Figure 7.30. You have the option to close the download screen or jump directly to the list of videos by selecting Play Now.

Selecting Play Now takes you to the Games folder in the Dashboard and directly opens the Game Trailers screen. It might make more sense to store movie trailers in the Media folder under Videos, but the Xbox 360 treats movie trailers just like game trailers, so they are listed together in the Games folder.

Figure 7.31 shows the Trailers screen, which shows the list of all the videos you've downloaded from Xbox Live. Selecting a trailer from the list brings up the trailer options screen, shown in Figure 7.32. In this screen, you can play the trailer or delete it.

> **Beyond the Manual**
>
> You can delete content that you've downloaded by using the System folder of the Dashboard.

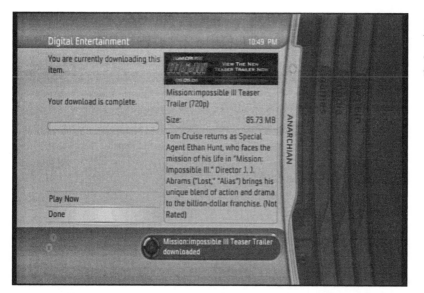

Figure 7.30
The movie trailer has been downloaded and is ready for viewing.

Figure 7.31
My extensive collection of movie trailers.

Beyond the Manual

I will show you how to make changes to your Xbox Live account in the next chapter.

Figure 7.32
You can delete the trailer
after playing it if you wish.

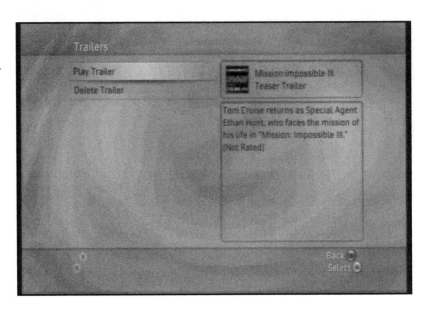

the gadget geek's guide to

Your Xbox 360

8

Xbox Live Marketplace: Game Demos, Trailers, Themes, and Microsoft Points

Xbox Live was launched in early 2002 to facilitate online gameplay for original Xbox games that were designed to take advantage of online multiplayer gameplay. Some of the early launch titles did not have online multiplayer support because Xbox Live was not ready for developers in time for the launch of the Xbox. That is why such popular games as *Halo* were fantastically popular in single-player, cooperative, and multisystem link (that is, LAN party) game modes but had no online capabilities. Several sequels to popular games, such as *Halo 2*, did provide online multiplayer modes.

Another good example is *Tom Clancy's Splinter Cell* (see Figure 8.1), an extraordinary single-player game that became playable online with the sequel, *Splinter Cell: Pandora Tomorrow* (see Figure 8.2). One could use the *Splinter Cell* series as a case example of how a game engine is improved with each new iteration of a game. The third game in the series, *Splinter Cell: Chaos Theory*, incorporated the compelling gameplay of the original single-player experience with the online capabilities of the *Pandora Tomorrow* so that it was possible to play cooperatively.

Figure 8.1
Tom Clancy's Splinter Cell.

I don't want to single out one development studio over another, but publisher Ubisoft does have some compelling games as a result of publishing for author Tom Clancy. The *Ghost Recon* and *Rainbow Six* series have been popular along with the *Splinter Cell* series, because they revolve around themes that are common in Tom Clancy's novels.

Figure 8.2
Splinter Cell: Pandora Tomorrow.

XBOX LIVE MARKETPLACE

Make sure you are connected online. This should have been taken care of automatically if your Xbox 360 is plugged into a network connection and you have already configured it for your network. To play games online, it goes without saying that your network must have an Internet gateway. As I explained in Chapter 7, "Going Online with Xbox Live," an Internet gateway is usually a router that hooks up to your cable or DSL modem, providing multiple network connections through the single "pipe" of your broadband connection. You can manually connect to Xbox Live by selecting the Gamertag option at the top of every folder in the Dashboard. Figure 8.3 shows the Xbox Live folder; note that I am already logged in to Xbox Live (as "ANARCHIAN").

Move the selector to the option called Xbox Live Marketplace and select it. This brings up the screen shown in Figure 8.4. Let's explore this screen because it is loaded with content. Notice the advertisement window on the right side of this screen. You can move the selector to this advertisement and select it if you are interested in ads that come up while browsing Xbox Live Marketplace. These ads should be relevant to the content on Xbox Live, because Microsoft does not operate Xbox based on advertising income (which is how sites like Yahoo and Google primarily derive their income). You have to pay for access to Xbox Live, so advertising is not a huge factor for Microsoft in this context. Therefore, you should not expect to see an ad here for nontargeted products like exercise equipment or magic weight-loss pills. You will, however, see ads here for new Xbox Live services and new Xbox 360 products.

Figure 8.3
The main Xbox Live folder in the Dashboard.

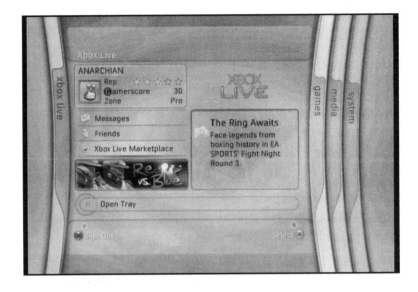

Figure 8.4
Xbox Live Marketplace.

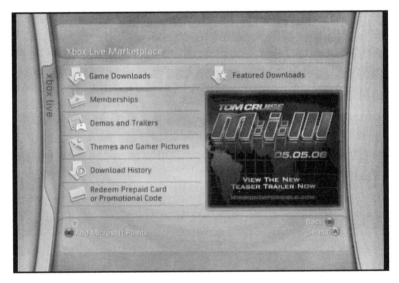

The following options are available on the Xbox Live Marketplace screen. (Note that these options are subject to change as Microsoft updates Live with new features.)

- Game Downloads
- Memberships
- Demos and Trailers

- Themes and Gamer Pictures
- Download History
- Redeem Prepaid Card or Promotional Code
- Featured Downloads

Let's explore these features, which make up the core of Xbox Live Marketplace.

Game Downloads

The Game Downloads option brings up the screen shown in Figure 8.5. This screen has two tabs: Played Games and All Games. The Played Games tab shows you the games you have played and any downloads that are available for these games.

Figure 8.5
Listing of game downloads in Xbox Live Marketplace.

Browsing Played Games

If you look at the list of games shown in Figure 8.5, you'll see that I've played *Joust*, *Gauntlet*, *Kameo*, *Perfect Dark Zero*, and *PGR3* (*Project Gotham Racing 3*). In some of these items, you can see that I have already downloaded some of the available content. For instance, the game *Joust*—which is an available download in Xbox Live Arcade, covered in the next chapter—has two available downloads. If you select the game, the screen shown in Figure 8.6 appears. This screen shows information about and available content for this game that you can download.

Figure 8.6
The *Joust* game has a free trial version and a full game available for download.

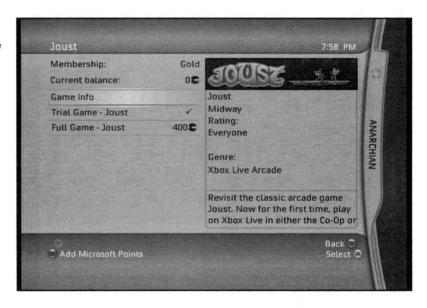

As you can see, the trial version of the game *Joust* is available as a free download. You can download the free trial to your Xbox 360's hard drive and then play the game. The full game is available for 400 Microsoft Points. I'll go over Microsoft Points later in this chapter.

Browsing All Games

From the Game Downloads screen, you can move the selector to the right to select the All Games tab. This screen, shown in Figure 8.7, provides three options:

- Alphabetical List of Games
- Xbox Live Arcade
- Games with New Downloads

The alphabetical list gives you a listing of every game. If you want to see just the list of downloadable "casual" games in Xbox Live Arcade, choose the second option. You can also select the third option if you have already browsed the games list extensively and just want to get an update of what's new. Select the first option to bring up the screen shown in Figure 8.8.

This list will definitely change over time as new content is added to the list, making online content available for existing games and with the addition of new games as they are released. You will also see the number of items shown for some games change as new content is added to them. Because this list is comprehensive, you will see the downloadable "casual" games listed with the retail games.

Figure 8.7
Browsing the entire list of games (played or not).

Figure 8.8
The complete list of downloadable games and game content.

Figure 8.9 shows an interesting selection. This is the available downloads for the game *Need for Speed: Most Wanted*. Look at the size of that game demo: almost a gig! You can expect to see many more game demos in the Marketplace available for download as time goes on because this is the easiest way to market a game. Unfortunately, there's no option here to unlock the full version of *Most Wanted* using the Marketplace. But wouldn't that be great? It may yet be a reality in due time.

Figure 8.9
The *Need for Speed: Most Wanted* game demo is almost a 1GB download!

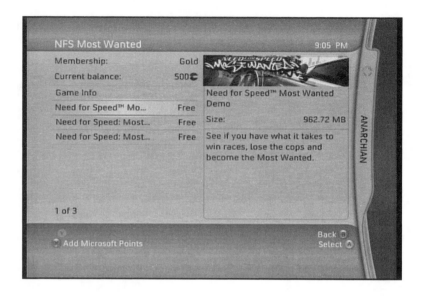

Let's go ahead and download the game demo now to see how fast we can fill up that 20GB hard drive. By selecting the game demo, you'll see the screen shown in Figure 8.10.

While the game demo is downloading, you see a progress bar (see Figure 8.11). Unfortunately, this progress bar doesn't show the estimated time to completion, which would be a welcome feature with large downloads like this one. When the download has completed, you see the screen shown in Figure 8.12.

Figure 8.10
The game demo is available as a free download.

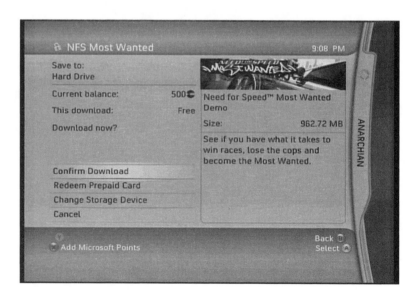

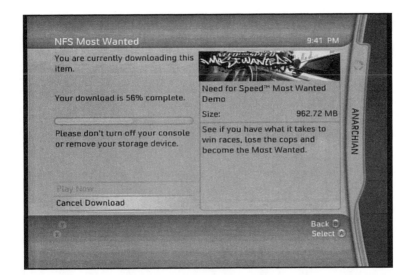

Figure 8.11
The game demo is downloading.

Figure 8.12
The download is complete; you may now play the game.

Memberships

Back at the main Xbox Live Marketplace screen, the second option takes you to the Memberships screen, shown in Figure 8.13. This screen shows you what membership you have subscribed to and what types of memberships are available. If you have a Monthly Gold Membership and you want to continue using Xbox Live beyond one month, you may want to upgrade to a more affordable membership option because it is cheaper when you subscribe for 3 or 12 months.

Figure 8.13
Viewing available Xbox
Live account membership
types.

Demos and Trailers

The main Xbox Live Marketplace screen has another menu item called Demos and Trailers that brings up the screen shown in Figure 8.14 when selected. This content is also available in the Game Downloads section of the Marketplace but is provided here along with movie trailers in a more convenient listing. (For instance, this list does not include gamer pictures or themes.)

Figure 8.14
The Demos and Trailers
screen.

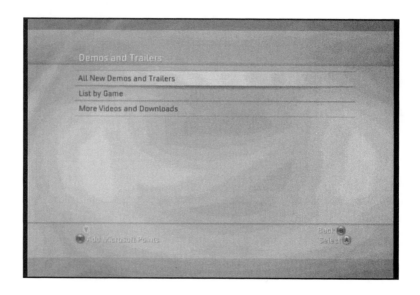

The first option, All New Demos and Trailers, lists only recently added game demos and game/movie trailers, bringing up the screen shown in Figure 8.15.

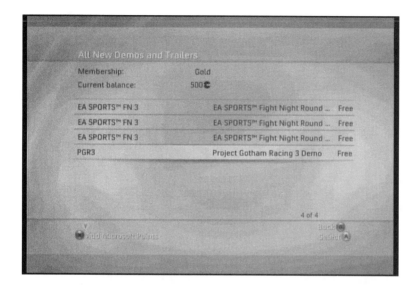

Figure 8.15
Listing of new demos and trailers.

The second option, List by Game, shows the entire list of game demos and trailers (see Figure 8.16). This is a comprehensive listing that you may want to use to locate an older demo or trailer.

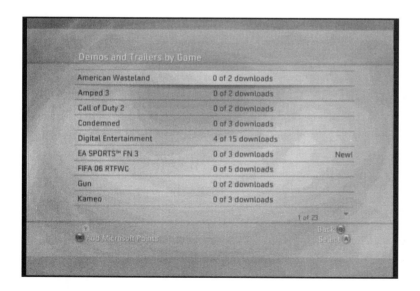

Figure 8.16
The complete listing of demos and trailers.

If you select More Videos and Downloads, you get a list of video trailers that you can download and watch on your Xbox 360. This includes trailers of video games and feature films. Figure 8.17 shows a trailer of the film *Aeon Flux*.

Figure 8.17
You can download movie trailers for free.

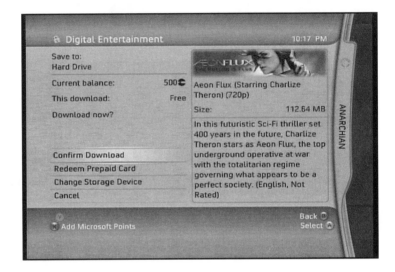

Themes and Gamer Pictures

The Themes and Gamer Pictures menu choice in Xbox Live Marketplace provides you with a convenient listing of just the Dashboard themes and gamer pictures available for download, shown in Figure 8.18.

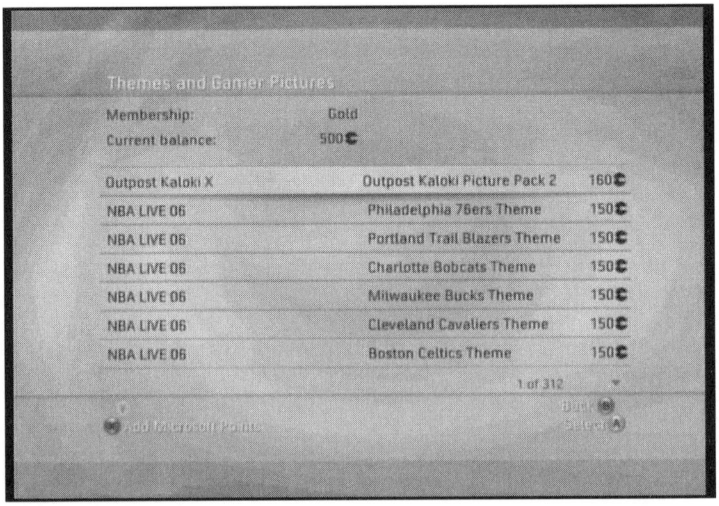

Figure 8.18
List of downloadable themes and gamer pictures (for a fee).

I have decided to download the *Quake 4* theme, shown in Figure 8.19. This theme will transform my Xbox 360 Dashboard from the standard to a skinned version of the Dashboard featuring the game.

Figure 8.19
Confirming the purchase and download of this new Dashboard theme.

Figure 8.20 shows the theme being downloaded, while Figure 8.21 shows the completion screen with instructions on how to use the newly acquired theme.

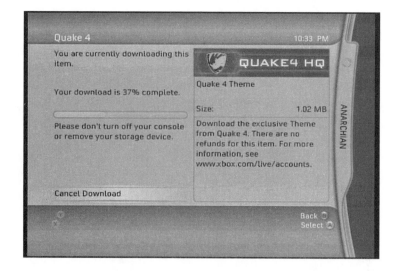

Figure 8.20
The *Quake 4* Dashboard theme is being downloaded.

Figure 8.21
The *Quake 4* theme has
successfully downloaded
and is ready to be used.

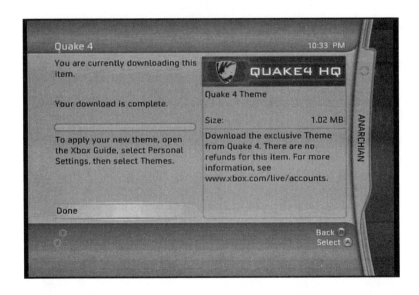

Changing the Dashboard Theme

Now that I've downloaded a new theme, I want to install it in my Xbox 360 Dashboard. I can
do this by using the Xbox Guide, which comes up by pressing the "jewel" at the top of the
controller or media remote. (It's silver with a green X over it—this is the official Xbox 360 logo.)
Select Personal Settings, as shown in Figure 8.22.

Figure 8.22
The Xbox Guide is used to
customize your Xbox 360
Dashboard.

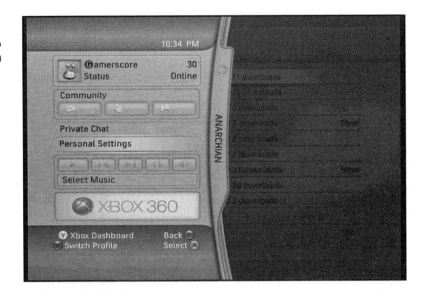

This option brings up the screen shown in Figure 8.23, which allows you to customize the Xbox 360 Dashboard with your preferences—including the theme you want to use.

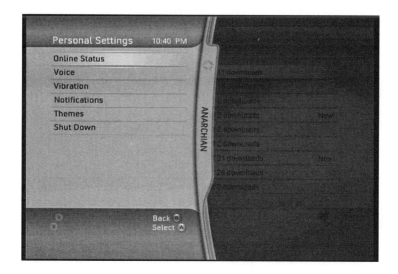

Figure 8.23
The Personal Settings screen in the Xbox Guide.

There are a lot of useful options here. You can explore all of them, but I want to bring your attention to the Themes option. Selecting it brings up the Themes screen shown in Figure 8.24.

When you select a particular theme, the entire Dashboard is transformed before your eyes. The background for each of the four Dashboard folders (Xbox Live, Games, Media, System) contains a unique bitmap image; the color scheme of the Dashboard changes; and the textures used for

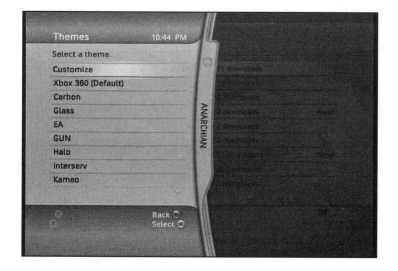

Figure 8.24
List of available Dashboard themes.

175

the Dashboard folders are "skinned" to the theme you've chosen. Figure 8.25 shows the Dashboard with the *PGR3* theme.

Figure 8.25
The Dashboard theme based on *Project Gotham Racing 3*.

Some of the themes shown in my list were available as free downloads, and some came preloaded with the Xbox 360. Still others, like the *Quake 4* theme, cost a few Microsoft Points to acquire. Figure 8.26 shows my favorite theme: *Halo*.

Figure 8.26
The Dashboard theme based on *Halo*.

Let's check out the theme I just purchased in the Marketplace. Figure 8.27 shows the *Quake 4* theme. This is a minimalist theme with barely visible background images in each folder and no skinning or color scheme. I feel like I just flushed 150 Points down the drain—especially when I compare this to some of the excellent free themes!

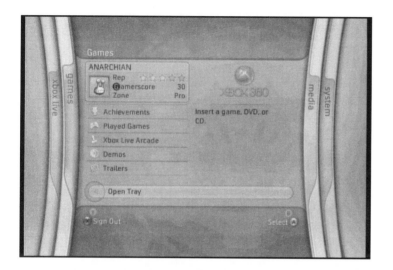

Figure 8.27
The Dashboard theme based on *Quake 4*.

Beyond the Manual

You can copy themes and other downloaded content to a memory card for transport to your friend's 360 and then use such items on other 360s, but there is a caveat: You must log into Xbox Live on that system to use the theme or other content item. Why? Because your downloaded content is tied to your Xbox Live account with a technology called Digital Rights Management (DRM)—a term you may be familiar with if you purchase music files online.

Meltdown

There are quite a few useless themes and gamer pictures for sale on Xbox Live Marketplace. Be careful that you don't spend too much money just to see what they look like, because some—like the *Quake 4* theme—are real stinkers!

Download History

The Xbox Live Marketplace includes an option called Download History that shows you a list of everything you have downloaded. This is a quick way to check out everything you've gobbled up while you've been a member of Xbox Live. See Figure 8.28.

Figure 8.28
The Download History list includes everything you have downloaded from Xbox Live.

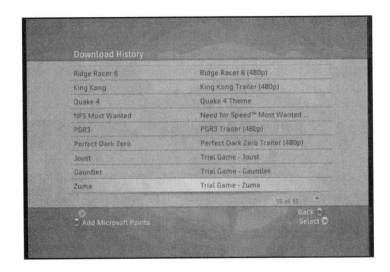

Redeem Prepaid Card or Promotional Code

The Redeem Prepaid Card or Promotional Code option allows you to use the virtual keyboard to punch in a special code to redeem Microsoft Points or Xbox Live membership credit.

Featured Downloads

The Featured Downloads item is located on the right of Xbox Live Marketplace above the advertisement window. This provides you with a list of what might be considered (subjectively) the best downloadable content available (see Figure 8.29).

Figure 8.29
The Featured Downloads in Xbox Live Marketplace.

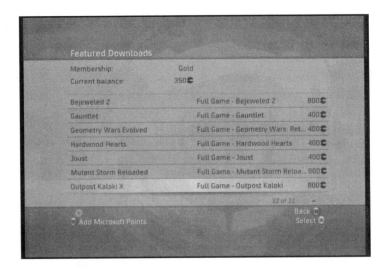

MICROSOFT POINTS

Microsoft designed Xbox Live Marketplace to function with its own internal economy. Rather than pay publishers directly for games you want to download and play on Xbox Live Arcade, Microsoft decided to abstract the concept of currency within Xbox Live so that you do not have to work with real money (which would probably be exchanged via credit card or debit card). One Microsoft Point is worth about 1.25 cents at the time of this writing.

This means that a quarter (25 cents) will get you 20 Microsoft Points. Or, more simply, one Microsoft Point costs $0.0125, or 1.24 cents. This exchange rate will likely remain constant over the coming years; what is likely to change is the price charged for downloadable content (such as Xbox Live Arcade games). It doesn't matter how many Points you buy, because no discount is offered for larger quantities. This exchange rate is simple enough. It does not correspond to an international currency exchange rate with the U.S. dollar, so its arbitrary amount online is the same regardless of real-world currency exchange rates. One U.S. dollar will buy 80 Points, whereas one UK pound will buy 120 Points. Whether dramatic changes in the real-world exchange values will cause Microsoft to adjust the price of Points is unknown at this time. This could certainly lead to a problem of currency speculation as traders buy and sell Points in different markets to make money. Figure 8.30 shows a typical prepaid Point card available in retail stores.

Figure 8.30
You can use a prepaid card to fill your Xbox Live account with 1600 Microsoft Points.

You can purchase Microsoft Points from almost every screen within Xbox Live Marketplace. For instance, in the screen shown in Figure 8.31, there's an option at the bottom (denoted with the blue X button) called Add Microsoft Points. Microsoft makes it easy to purchase more Microsoft Points. If you are a paying subscriber to Xbox Live, you have the option to use your current payment method to purchase Microsoft Points.

Figure 8.31

Listing of available content for purchase on Xbox Live. (Note the option Add Microsoft Points.)

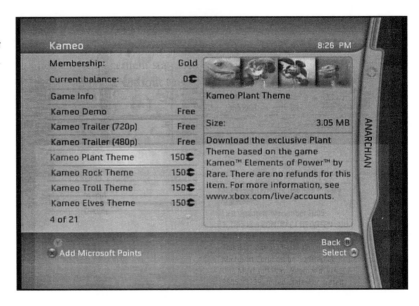

Pressing X on your controller or media remote brings up the Add Microsoft Points screen shown in Figure 8.32.

Figure 8.32

You can purchase Microsoft Points using this interface.

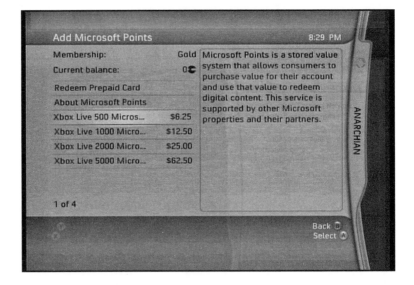

If you do the math (see Table 8.1), you'll learn that it makes no difference how many Points you buy; they cost the same regardless of quantity. Therefore, you might think of Points more as a currency exchange rather than a product sold in units.

Table 8.1 Purchasing Microsoft Points

Number of Points	Cost
500	$6.25
1,000	$12.50
2,000	$25.00
5,000	$62.50

In Figure 8.33, I have chosen to buy 500 Microsoft Points, which will be added to my Xbox Live account. I already have one payment method set up for my Xbox Live membership fee, so that is used by default for the purchase of the Points. I can change the payment option to choose a different credit card or to deposit the Points from a prepaid card, or I can choose Confirm Purchase.

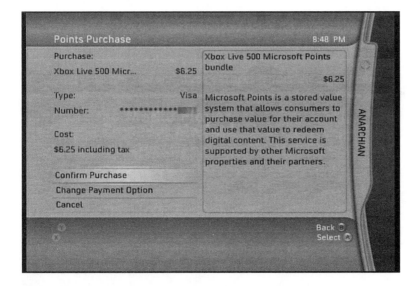

Figure 8.33
Getting ready to purchase 500 Microsoft Points using a credit card.

Selecting Confirm Purchase commits the transaction and charges my credit card (or other payment device) to cover the cost of the Points. You can see the confirmation of the completed transaction in Figure 8.34.

After closing this screen, I am returned to the previous screen that I was viewing when I pressed X to purchase some Microsoft Points. You can add more Points to your account at

Figure 8.34

Transaction complete; I have just purchased 500 Microsoft Points.

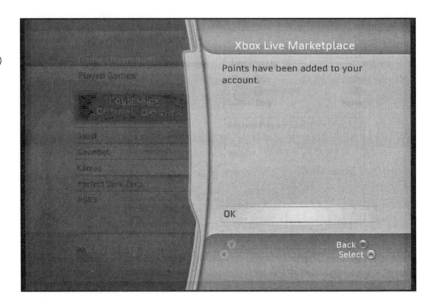

any time without losing your current position within Xbox Live Marketplace. As you can see in Figure 8.35, I now have a current balance of 500 Microsoft Points in my Xbox Live account.

Figure 8.35

My Xbox Live account now has 500 Points available for online purchases.

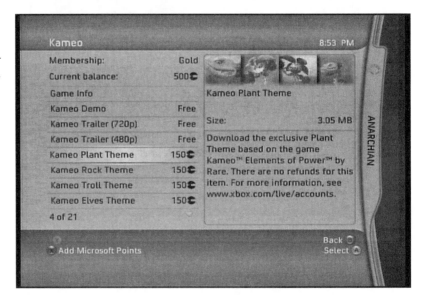

9

Xbox Live Arcade: Downloading "Casual" Games

The term *casual game* once referred to a segment of the video game retail market that included substandard games sold in the budget aisles of Wal-Mart and similar stores. Today, the casual game movement, as you might refer to it, has turned a new page in life. Casual games are now a respected genre in the video game industry, and many publishers are getting in on this lucrative multibillion dollar market by releasing smaller games through various mediums—such as Xbox Live Arcade.

CASUAL GAMES

The most common form of distribution of casual games has, up to this point, been through the Web, with several Web-based distribution companies earning a sizeable profit from downloadable casual games. One of the most successful examples includes RealArcade (http://www.realarcade.com), which offers many of the industry-leading casual games such as *Zuma Deluxe* (which is now available on Xbox Live Arcade).

The casual game segment of the video game industry is now taken seriously and is significantly represented at the major conferences, including the following:

- Electronic Entertainment Expo (http://www.e3expo.com)
- Game Developers Conference (http://www.gdconf.com)
- Austin Game Conference (http://www.gameconference.com)
- Casual Games Conference (http://www.casualgamesconference.com)

At these conferences, you will see lectures about the development and distribution of casual games over the Web, through retail shelves, and through online sources like Xbox Live Arcade. At the Austin Game Conference in 2005, I attended several sessions devoted entirely to casual games. Most of the expo floor seemed to be dedicated to this once-niche subject, with companies like Microsoft being represented solely through its new Microsoft Casual Games studio (which was promoting casual game development for MSN Messenger and Xbox Live Arcade).

In the real world (something game publishing companies seem to be oblivious to), casual games have been around for many years, and to some who have been in the casual game business for a long time, this latest trend is indicative of the cluelessness that abounds in many corporate board rooms when it comes to game development. Marketing should not be driving games, but that is an unfortunate side effect when big bucks are involved. That marketing people missed the casual game trend when it first started picking up pace in 2002–2003 should tip off the smaller game developers to the fact that most large game publishing companies have no idea how to make predictions in this market and are simply responding to each trend.

Predicting Best-Selling Games

Gamers define what games are popular. This fact has been true for the past 20 years, but some people simply ignore that as the most significant trend to follow. Gamers decided that real-time strategy games were getting old several years before the publishers figured that the trend had passed, and scores of real-time strategy game projects were cancelled in mid-development—or were delayed for a long time—as a result. Just compare the sales figures of recent RTS games such as *Homeworld 2*, *Age of Empires III*, and *Empire Earth II* with the previous games in these series. You'll find that gamers were enamored with real-time strategy for a long time, but that fad has passed for the majority of gamers.

The latest trend now is the MMOG—Massively Multiplayer Online Game, dominated by the subcategory MMORPG—Massively Multiplayer Online Role Playing Game. There are so many MMOGs in development today that I've lost count of them all. What happens in marketing and sales is that a ridiculously successful game franchise is used as an example for a genre, and then new contracts are signed to "cash in" on the trend set by a successful example. The most obvious example right now is Blizzard's *World of Warcraft*. This is an extraordinary game that has shattered all expectations and pushed the bar for this genre so high that it is doubtful that any other company will match *World of Warcraft* for many years. Such was the case with Blizzard's previous trend-setting games such as *Warcraft II*, *Warcraft III*, *Diablo II*, and *StarCraft*. This company knows how to make games, plain and simple—it's not about capitalizing on a trend with Blizzard; it's about creating trends.

What game publishing companies need to do is start recruiting the management from game studios and hardcore gamers rather than business professionals drawn from wholesale and retail sales management positions, because they simply don't understand the market. How do you expect someone to sell cars if he has never before driven a car? This describes the opinion of many gamers who are browsing the aisles at a local retail or online video game store: a pint of apathy with a shot of sarcastic humor. Gamers are not your typical, predictable consumer. They have experienced the greatness of games like the *DOOM* and *Quake* series, *Battlefield 1942* and *Battlefield 2*, and *Warcraft III* and *World of Warcraft*. These are genre-busting, trend-setting powerhouse games that are successful because they are ridiculously fun to play, not because they have all the ingredients of some marketing company's feasibility study.

Marketing people do fulfill an important role for game publishing companies, but their approaches to predicting trends might be described as the shotgun approach, where games are published to appeal to as wide an audience as possible; in the process, such games do not appeal to most gamers. Wouldn't it be easier to just ask people what they want? Better yet, why not hire a gamer to design new games based on fun gameplay? Few marketing professionals would have predicted that Nintendo's weird *Super Smash Bros. Melee* game would outsell *Age of Empires III*. But a gamer might have made that prediction after playing both games for an hour. *Melee* is fast-paced and silly, but it's fun and social with four-player support on a single television. *Empires III* is hard-core, serious, difficult, and time consuming, with multiplayer network support. Who but a hardcore gamer would know the difference?

The big game publishers should recruit more experienced game players and designers into management roles. But they aren't likely to do so. The automobile industry knows from experience that when you want to build a good car, you bring drivers in to help design cars that will excel in their niche target audiences. Left solely to marketing, all cars would look alike—they would all resemble SUVs and minivans. This actual trend left many automobile manufacturers in dire straits in recent years. What do we see now? Specialty cars are again being prototyped and shown at national events (such as the Detroit Auto Show) to galvanize the public's interest in "cool" cars once again. Now we see the return of the muscle car (circa 1970) with new designs like the reinvented Ford Mustang, Dodge Challenger, Chevy Camaro, and Shelby GT-500. Who demands performance and fun in automobiles? Consumers! Only after realizing the folly of selling to a poll group rather than to a real customer have the domestic automakers in the United States begun to turn around their weak sales.

Nintendo, for example, has a huge lead in the casual game market because this company has been producing casual games for two decades, and it is Nintendo's staple. Just take a look at *Super Smash Bros. Melee* for the Nintendo GameCube as an example (see Figure 9.1). This game supports one to four players simultaneously and features action-packed fighting using Nintendo's assortment of mascot characters, such as Mario, Bowser, Kirby, Seamus, and so on. Who would have suspected this game would be a hit? Only just about everyone who has ever played it! The success of the Nintendo 64 version of *Smash Bros.* and other oddities like *Conker's Bad Fur Day* should tip off game publishers to the types of games people want to play.

Figure 9.1
Super Smash Bros. Melee is a popular casual game for the Nintendo GameCube.

I suspect that the marginal sales of even the weakest games may be enough to warrant their development. But the real danger in that business plan is that resources are tied up reloading the proverbial shotguns when precise shots would be more effective. For examples of fun gameplay and expert polish and presentation, take a look at any game that Blizzard has created in the past decade for inspiration.

Casual Hype

Given all the hype and hustle and bustle surrounding casual games, it might help to define exactly what attributes determine whether a game can be categorized as casual. First of all, what types of games can you think of that are *not* casual in nature? I can think of hundreds, because most games have a large scope with great depth. (For example, Blizzard's *World of Warcraft* is about as far away from a casual game as it is possible to get, given its great complexity.)

A casual game is usually defined as a game that you can play in five minutes or less, and it is usually assumed that you do not need to download or install a casual game before playing it, but that is not a strict requirement. (You must download Xbox Live Arcade games before playing them on your Xbox 360.) Another assumed attribute of a casual game is that it does not require instructions to play. Therefore, any casual game that adds enough complexity so as to require instructions may not be a casual game.

I define a casual game as a game that I can quickly and easily play for 5–10 minutes. Most casual games fall into this category, but some are a little difficult to figure out without help. By this definition, most of the old "coin-op" arcade games fall into the category of casual games.

DOWNLOADING CASUAL GAMES

You need an active Xbox Live account to download Xbox Live Arcade games to your Xbox 360. After you're logged in, go to the Games folder in your Xbox 360's Dashboard and select Xbox Live Arcade, as shown in Figure 9.2.

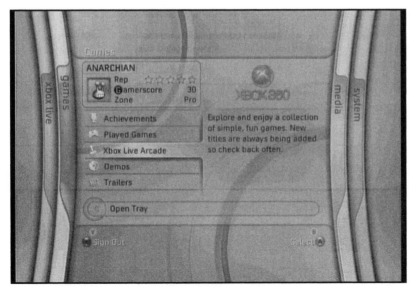

Figure 9.2
The Games folder in the Dashboard.

This brings up the Xbox Live Arcade screen, shown in Figure 9.3. From this screen, you can bring up a list of games you have installed on your system and download new games (either the trial versions or full versions). I recommend downloading everything you find on Xbox Live at least once just to try out the many different games that are available. You can find many hidden gems if you just give these games a try.

Figure 9.3
Xbox Live Arcade.

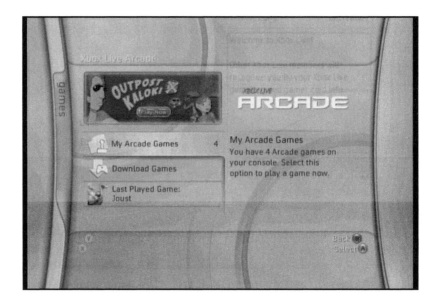

Your Arcade Games

Selecting My Arcade Games brings up a screen with the same title, shown in Figure 9.4. This screen shows a list of all games you have already downloaded from Xbox Live Arcade and which are ready to be played. This list shows that I have already downloaded the trial versions of *Gauntlet, Hexic HD, Joust,* and *Zuma.* This is just a sampling of the many games available to you on Xbox Live Arcade. In fact, I predict that you will spend more time playing these casual games than you will spend playing the expensive retail games! Again, it's about convenience and time. A casual game requires a small time investment to learn and play, and you can easily jump into a casual game whenever you have a few minutes to spare and then quit whenever you want.

By selecting a game from the list, you bring up a detail screen about the game, showing whether you have the trial or full version. There's an option to delete the game, which is useful if you no longer play that game. I go into more detail about this screen in the later section titled "Playing Casual Games."

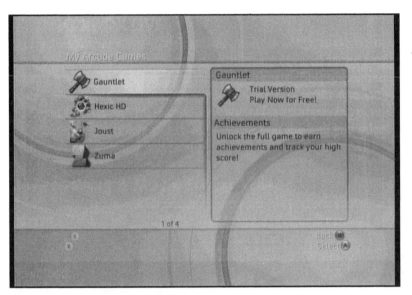

Figure 9.4
The My Arcade Games screen.

Beyond the Manual

Why do you suppose most retail games seem to require more of an investment of your time compared to casual games? By the very definition, a casual game is a game you can play any time you have a few minutes to spare. In contrast, most retail games require a more significant time investment just to fire up the game—for instance, due to title screens, copyright screens, opening videos, and so forth. Perhaps casual games are appealing simply because they are not so *involved*. A simpler analysis might describe retail games as heavyweights, and casual games as lightweights.

Downloading Games

If you have never browsed Xbox Live Arcade, your games list might be empty (although I believe *Hexic HD* is installed by default on the hard drive). Due to the size of most of the games on Xbox Live Arcade, you probably will not be able to download and play games using a memory unit (with only 64MB), although that may be possible for some of the smaller games. From the main Xbox Live Arcade screen, choose the option called Download Games, as shown in Figure 9.5.

This brings up a list of games sorted by genre, as you can see in Figure 9.6. Selecting a specific genre, such as action, brings up a list of games within that genre. Figure 9.7 shows several games in the list of action games.

Figure 9.5
Xbox Live Arcade.

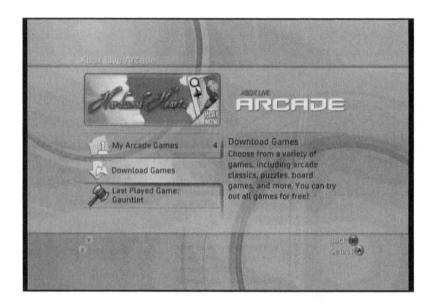

Figure 9.6
Xbox Live Arcade games,
sorted by genre.

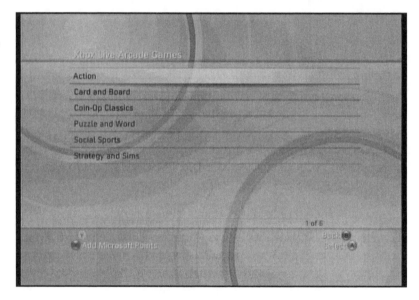

Let's download a game to see how this process works. I will download the game *Mutant Storm Reloaded*, as shown in Figure 9.8.

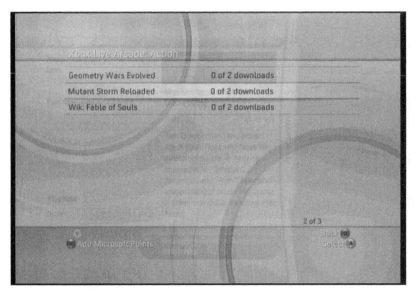

Figure 9.7
Action games.

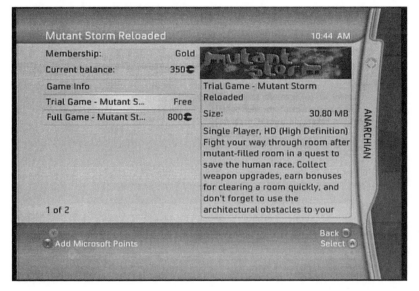

Figure 9.8
The download screen for *Mutant Storm Reloaded*.

Selecting the Trial Game option for this game brings up the download screen shown in Figure 9.9. This screen looks the same even if you are buying a full version of a game using Microsoft Points. In this case, however, the cost is free.

Figure 9.9
Confirming the download
of this game.

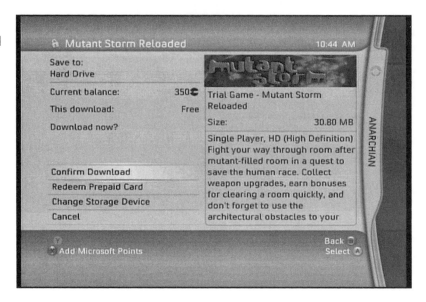

When you confirm the download, the download commences, and you can watch its progress, as shown in Figure 9.10. When the download completes, you see the screen shown in Figure 9.11, with an option to play the game immediately.

Figure 9.10
Downloading the *Mutant Storm Reloaded* game demo.

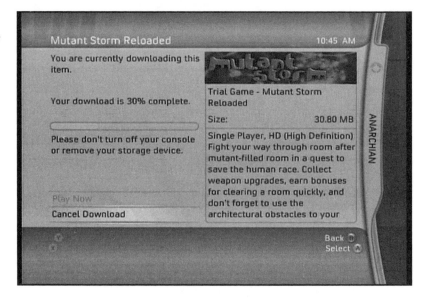

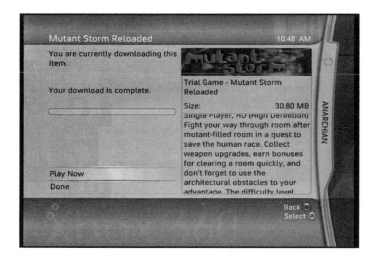

Figure 9.11
The download has completed, and the game is ready to be played.

Purchasing Games

You can only have so much fun with the trial versions of the Xbox Live Arcade games, because they are purposely limited in gameplay to give you just a taste for what the game is like. To really enjoy the full potential of these games, you must purchase them using Microsoft Points.

After playing numerous casual games that I've downloaded from Xbox Live Arcade, my first choice for purchase is *Zuma*. I really enjoy this game and think it represents the genre perfectly. It's easy to play, addictive, and requires no instructions. Figure 9.12 shows the purchase screen for *Zuma*. This game costs 800 Microsoft Points. I only have 350 in my account, so when I

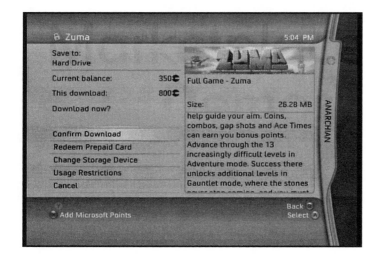

Figure 9.12
Purchasing a game with Microsoft Points.

confirm this purchase, the screen shown in Figure 9.13 comes up, telling me that I have insufficient funds. Then it gives me an opportunity to purchase more Microsoft Points.

Figure 9.13
Acquiring more Microsoft
Points to make a
purchase.

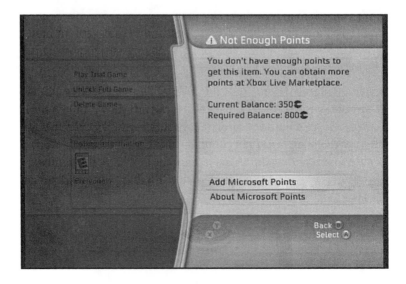

When I select Add Microsoft Points, the screen shown in Figure 9.14 appears. As I explained in Chapter 8, "Xbox Live Marketplace: Game Demos, Trailers, Themes, and Microsoft Points," it doesn't make a difference how many Points you choose, because there's no discount for buying

Figure 9.14
Purchase options for
Microsoft Points.

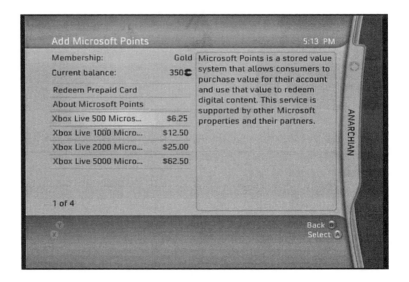

more Points. I recommend just buying as many Points as you need at a particular time. In this case, I'm going to buy 500 Points.

After the transaction is complete, the screen shown in Figure 9.15 comes up again. This time, you can see that I have enough Points in my account to upgrade to the full version of *Zuma*, but it will use up almost all the Points in my account. If you want to determine how much something costs in real-world dollars and cents, just multiply the number of Points that an item costs by 0.0125. (Each Point costs one and a quarter cent.) The 800 Points for *Zuma* comes to $10. That's not a bad price for a fun game that I plan to play a lot. It will be even more fun with the full version's online features.

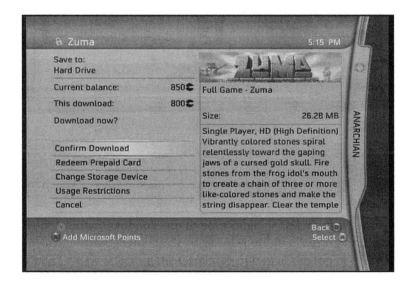

Figure 9.15
Upgrading to the full version of *Zuma* costs 800 Microsoft Points.

Now that I've purchased the full version of *Zuma*, the description of the game in My Arcade Games has changed to show my achievements and Gamerscore, shown in Figure 9.16. This game is so much fun that I'm getting addicted to it!

Beyond the Manual

When you purchase an item on Xbox Live, the item is listed in the inventory of your account, regardless of whether you have downloaded it. You can delete a game from your Xbox 360 and then download it again without having to pay for it a second time. You can also take your Xbox Live account on a memory unit to a friend's house and use it to download your purchased items again—but your friend will not be able to play those games when you leave.

Figure 9.16

The Gamerscore and achievements are now shown for the full version of *Zuma*.

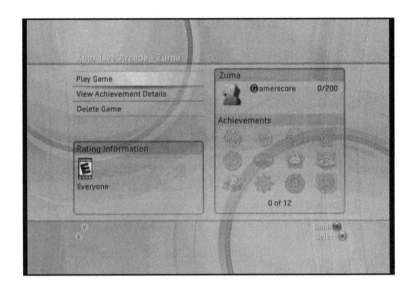

PLAYING CASUAL GAMES

I recommend that you download all the trial edition games that you can find on Xbox Live Arcade. Some apparently unassuming games are actually extremely fun games that you might not otherwise consider.

Gauntlet

Let's take *Gauntlet* for a spin. Select it from the list of downloaded games to bring up the detail screen, shown in Figure 9.17. This is an accurate port of the classic arcade game of the same name.

Geek Speak

Porting describes the process of converting a game from one platform to another. For instance, *Tom Clancy's Splinter Cell* was originally created for the original Xbox and then "ported" to the Nintendo GameCube and Sony PlayStation 2. Many games today are developed for two or three platforms simultaneously; this is called *cross-platform development* rather than porting.

The title screen for *Gauntlet* is shown in Figure 9.18. This screen is typical of all the Xbox Live Arcade games, giving you an option to play the game, to play online with others, to view the high score list (Leaderboards), and so on.

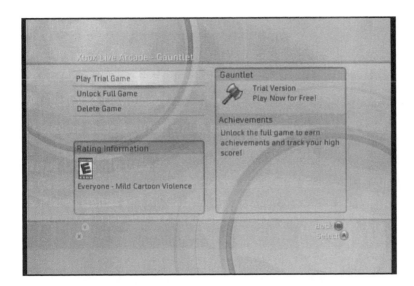

Figure 9.17
The detail screen for a downloadable game.

Figure 9.18
The title screen for *Gauntlet*.

Although you cannot *upload* your high scores to Xbox Live unless you purchase the full version of the game, you can *view* the high score list in some games. The high score list for *Gauntlet* is shown in Figure 9.19. Because you have an opportunity to see your name in the spotlight by scoring high on the high score list, there is considerable motivation to play the game and beat everyone else's score!

Figure 9.19
The "global" high score list for *Gauntlet* is stored on Xbox Live and shared with everyone.

The Achievements option in the main menu of *Gauntlet* brings up the screen shown in Figure 9.20. As you can see, I have not played the game very much because I haven't unlocked extra features yet. You can play the full version of *Gauntlet* online with up to three friends in cooperative mode. This feature alone is worth the cost of the game, because some of the unlockable features, such as Anti-Thief, work only in cooperative mode.

Figure 9.20
The list of achievements (unlocked items) in *Gauntlet*.

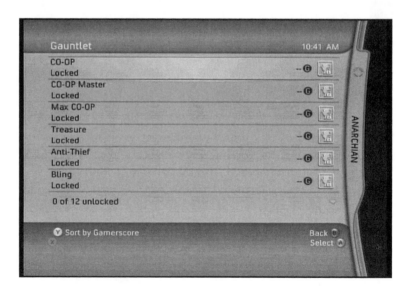

As you can see in Figure 9.21, this is an accurate translation of the original *Gauntlet* coin-op arcade game. This version has been enhanced from the original with multiplayer support over Xbox Live. Now you can play with up to three friends in co-op mode!

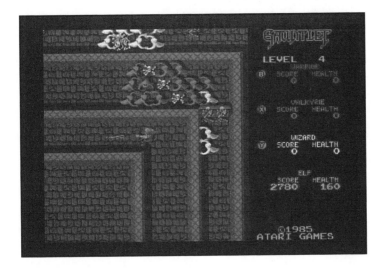

Figure 9.21
Gauntlet looks like the original coin-op, but it actually has a lot of new features.

Joust

Another classic coin-op arcade game that you can play on Xbox Live Arcade is *Joust*. The title screen is shown in Figure 9.22. You can play this game in arcade (single-player) mode, in versus mode (which requires two controllers), or on Xbox Live against other players around the world.

Figure 9.22
The main menu for *Joust.*

Joust features characters that fly on ostriches and wield spears with which to kill each other. You must be positioned higher than the enemy players to hit them with your spear; otherwise, they kill you. This game is a lot of fun in two-player co-op mode. See Figure 9.23.

Figure 9.23
Joust's graphics might look primitive, but the gameplay has survived the test of time.

Mutant Storm Reloaded

Mutant Storm Reloaded is a fun game with pretty graphics. It involves flying some sort of (mutant?) character around the screen—which seems to be in outer space—while shooting enemies (mutants?). Figure 9.24 shows the title screen. The gameplay is fun, but the design doesn't seem to fit the game's title. Figure 9.25 shows the gameplay.

Figure 9.24
The main menu of *Mutant Storm Reloaded*.

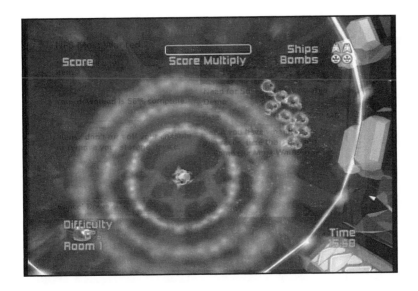

Figure 9.25
Mutant Storm Reloaded is an intriguing game that shares some gameplay elements with *Smash TV* and *Geometry Wars*.

Hexic HD

Hexic HD is a puzzle game that involves a hexagonal playing board with hexagonal jewels that you must line up in groups of three to eliminate jewels from the board. Figure 9.26 shows the title screen. To win the game, you must get as many points as possible from each combination. You must achieve a certain number of combos to move to the next level. Figure 9.27 shows the game screen.

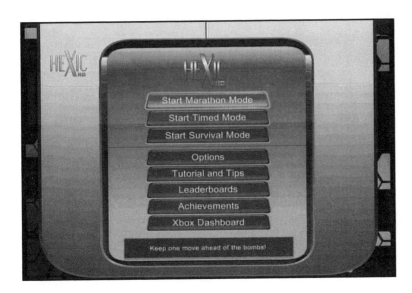

Figure 9.26
The main menu of *Hexic HD*.

Figure 9.27
Gameplay in *Hexic HD* can be challenging.

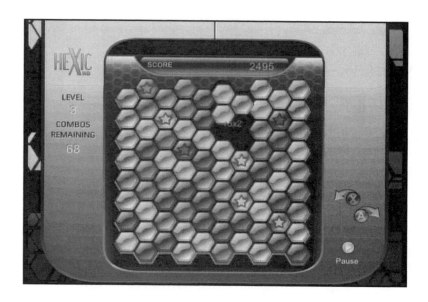

Zuma

Zuma is a challenging and fun game that involves shooting balls from a frog's mouth at a line of balls that are rolling around toward the center of the screen. The goal is to prevent the balls from reaching the center by matching three or more balls of like color to eliminate them from the line. Figure 9.28 shows the title screen of *Zuma*, and Figure 9.29 shows the game in action. This game is so much fun that it is the first game I purchased after playing the trial version.

Figure 9.28
The main menu of *Zuma*.

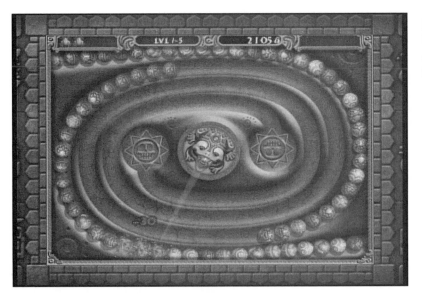

Figure 9.29
Here I have eliminated some balls by combining three like colors.

Geometry Wars

Returning to the My Arcade Games screen, you can see that I have downloaded several more games, including *Geometry Wars* (see Figure 9.30). The main menu of *Geometry Wars* is shown in Figure 9.31.

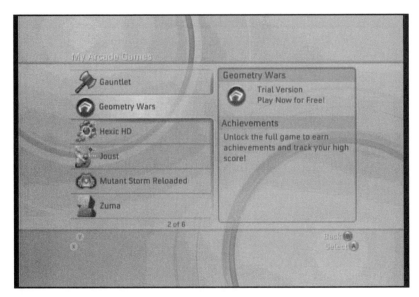

Figure 9.30
The list of games that I have downloaded.

Figure 9.31
The main menu of
Geometry Wars.

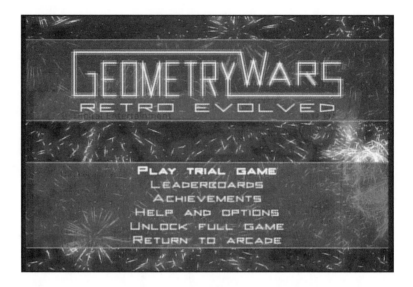

Geometry Wars was created by the team responsible for *Project Gotham Racing 3*. This arcade game actually started life as a mini-game inside *Gotham 2* and has since been made into a full-blown game that stands on its own now. The action in *Geometry Wars* is frenetic and requires a lot of practice to beat the higher levels. The most interesting aspect of the game is that there are black holes that seem to warp space around them, drawing in enemy objects and your ship. In fact, physics is a big part of the game. You will gain powerful weapons that warp space as they are fired, which produces a really interesting effect (see Figure 9.32).

Figure 9.32
The gameplay in
Geometry Wars is fast
paced.

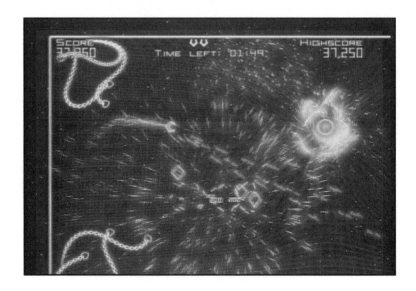

Smash TV

Smash TV was a classic coin-op arcade game. The game is a type of television game show, where the object is to shoot and otherwise decimate mutants to clear rooms and make your way toward the final room of the game, where you fight a boss character. The gameplay is frantic, but it helps that you can play co-op with a friend. Figure 9.33 shows the title screen, and Figure 9.34 shows the game in action.

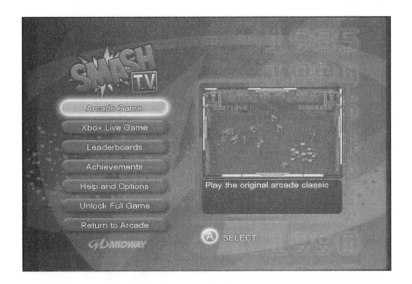

Figure 9.33
The main menu of *Smash TV*.

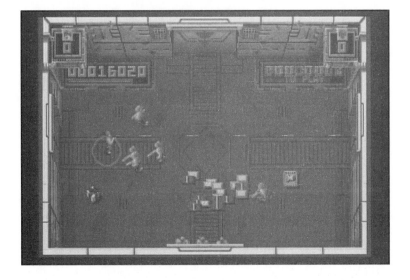

Figure 9.34
Smash TV was controversial in its time for being too violent.

the gadget geek's guide to

Your Xbox 360

10

Xbox Live Community: Ranking, Reputation, Voice Chat, Friend Lists, and Instant Messaging

Xbox Live is much more than just an online game server or network for hooking up with other players. When you sign up to Xbox Live, you gain access to a global game network with distinctive social features available for use in communicating with other players online. Xbox Live Arcade brings casual gaming to a much larger audience than is common with the high-profile games (such as id Software's *Quake 4*) by allowing casual gamers to play with each other through the medium of casual games (as discussed in the previous chapter). That's why it's obvious that this communication medium needs a rich set of tools to make it possible for gamers to communicate with instant messaging, to keep track of their online friends, and to talk with each other using the voice chat headset.

GAMERSCORE AND ACHIEVEMENTS

Let's start with the most obvious aspects of Xbox Live's social features. When you power up your Xbox 360, the Dashboard comes up with a focus on the Xbox Live folder, shown in Figure 10.1. This folder is displayed first because it is most likely where you will spend most of your time: sending messages to friends, initiating voice chat and instant messaging, and, of course, viewing your achievements.

Figure 10.1

The Xbox Live folder in the Dashboard.

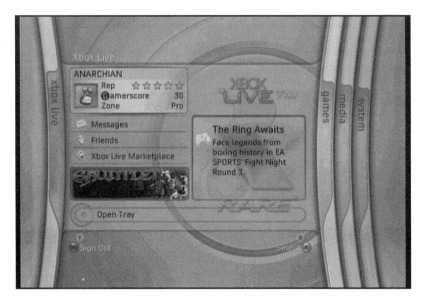

Geek Speak

The concept of game *achievements* is new for the Xbox 360 and is not applicable when playing original Xbox games. When you complete a level, win a race, or defeat a certain number of enemies in a game (just a few examples), you earn an achievement in that game. The developers who created the games you play can take advantage of a new feature in the Xbox 360 that allows them to save global achievements for a game that can be viewed in your Xbox 360 gamer profile. Each achievement is worth a certain number of points that are totaled to come up with your total Gamerscore.

Let's select the Xbox Live account by highlighting the account item at the top of the screen. This is called your *Gamercard*, and it is unique to you. In fact, you can share your Xbox 360 with your relatives and friends without affecting your own personal gamer profile—preserving your achievements and score in each game. Selecting your Gamercard brings up the Gamer Profile screen shown in Figure 10.2. You can use this screen to view your stats (such as your Gamerscore) and to make changes to your gamer profile. These are new features that were not available in the original Xbox.

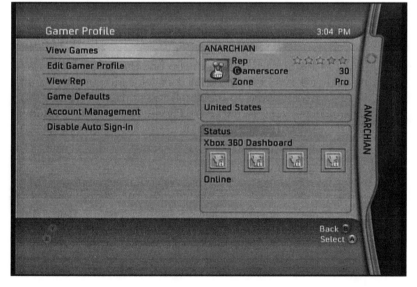

Figure 10.2
The Gamer Profile screen.

Your gamer profile is tracked with your Gamertag on your local Xbox 360. This information is shared with your Xbox Live account, so your achievement score is the same for both online and offline (single-player) games. The Gamer Profile screen has the following options that I will go over with you, one at a time.

View Games

The View Games option brings up the screen shown in Figure 10.3. This screen shows a list of all the games you've played along with your Gamerscore. I downloaded most of these games recently, and I don't own the full version. Because you can only accrue a score in a downloaded game by buying the full version, these games always show a zero Gamerscore until you purchase them. This is a harmless form of viral marketing that is comparable to the shareware business model. The shareware model was a popular way to sell PC games throughout the 1990s. It's the method used by id Software to originally sell *Wolfenstein 3D, DOOM,* and *DOOM II.*

Geek Speak

Your *Gamerscore* is the sum of all achievement points you have earned in all the games you have played, both online and offline.

The shareware model actually involved giving away games for free. However, as is usually the case, there is a catch: The free version of a shareware game is limited in gameplay, features, and scope. Some classic shareware games from the 1990s included *Jill of the Jungle* and *Kiloblaster* (by Epic Games), *Commander Keen* (by Apogee and id Software), and one of the earliest shareware games, *Captain Comic* (by Michael Denio).

Beyond the Manual

For an incredible nostalgic journey through the annuls of video game history, visit Moby Games at http://www.mobygames.com. I've spent many hours wandering through their incredible database of video games (with screen shots). If you would like to actually *play* many of the classic PC games featured at Moby Games, visit http://www.dosgames.com.

My score in some of the retail games is quite shameful, I'm loathe to admit. I do have an excuse: I've been *writing about* the Xbox 360 more than I've been *playing* it! This aspect of the Xbox 360 is probably the most enjoyable part of playing games, because everyone can see how good you are by looking at your Gamerscore (which is shared with other players who see you online). My Gamerscore of 30 reflects that I am a *newbie*, having not played very much yet. It's interesting to speculate how much of an impact the concept of an achievement score will have on the motivation of some gamers who will be compelled to beat their friends' scores. This applies not just to online gameplay on Xbox Live, but also to offline (solo) games, which still generate achievement points.

Geek Speak

A *newbie*, also spelled noob or n00b, is someone who has just started doing something for the first time and doesn't do it very well. A newbie might be thought of as an infant to a certain technology, without even the benefit of related experience. Geeks are notoriously harsh toward newbies, so the tendency is to pose or feign ignorance to avoid a beat-down. Be nice to newbies; we were all there once! Even Bill Gates was a newbie at one time—before he founded Microsoft at the age of 10 (or thereabouts).

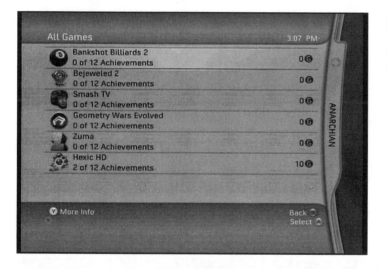

Figure 10.3
The All Games screen shows a list of games you have played.

Beyond the Manual

Your Gamerscore is the total of all your achievements in all the Xbox 360 games you have played. When you unlock a new level, discover a powerful weapon, or defeat a difficult "boss" character, these achievements are tracked with your Gamertag. In the past, accomplishments in games allowed you to progress within the game or to become a better player by acquiring special items. On the Xbox 360, those achievements do that, but they are also listed in your *gaming résumé*, so to speak (referring to your reputation).

While you are scrolling through the list of games you've played, you can select a game to bring up a list of achievements that you can complete in the game. Figure 10.4 shows the achievements for *Project Gotham Racing 3*. If you have spent even a moderate amount of time in any game—such as a few hours—you will quickly accumulate a good Gamerscore of several hundred points. I played *Kameo* for about two weeks right after it was released and racked up all kinds of achievements. Then I deleted my Gamertag and created a new one to see how the 360 keeps track of gamer profile data, and I lost those achievements. (I suffered for the interest of scientific research so you wouldn't have to!) This is an important consideration to make, because the Xbox 360 is capable of handling several players.

Figure 10.4

List of possible achievements for *PGR3*.

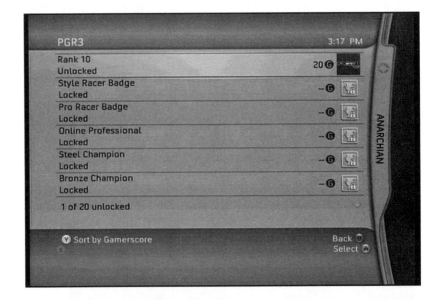

Meltdown

Do you want to protect your Gamertag and Gamerscore? If so, make sure anyone who shares your Xbox 360 creates his own Gamer account. Sign out before shutting off the Xbox 360 so that the next player has to sign on using his own Gamer account. It's similar to sharing a Windows XP computer and saving your own settings.

Edit Gamer Profile

Now let's look at your gamer profile. Figure 10.5 shows the Edit Gamer Profile screen. This screen has several options that allow you to customize your gamer profile. Customize your personal gamer profile to suit your style by selecting a picture to represent you when chatting with your friends online and setting your motto to a witty saying.

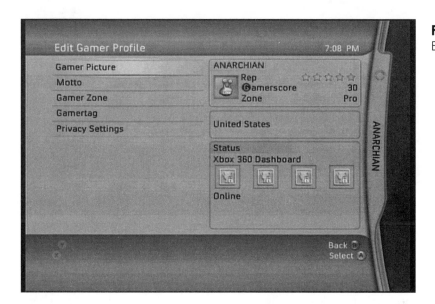

Figure 10.5
Edit Gamer Profile screen.

Gamer Picture

The Gamer Picture option brings up the Change Gamer Picture screen shown in Figure 10.6. This picture is an icon that is displayed along with your Gamertag to other players on Xbox Live.

Motto

The Motto option allows you to change your motto by typing in a short message (up to 21 characters maximum). Why would you want to use a motto? Because the length is so limited, there's hardly enough characters for even something simple like a Web site address, but if you have a Web site with a relatively short address, that might be a fun thing to enter here. For a while, my motto was "I'm Vin Diesel," but then I remembered hearing that he's actually a big-time gamer and didn't want to tick him off. So I changed it to "www.jharbour.com."

Figure 10.6
Changing your gamer profile picture.

Gamer Zone

The Xbox Live Gamer Zones screen allows you to change your preferred gamer zone in which you play games. Your zone determines the types of players that you are matched with on Xbox Live and is meant to assist with skill balancing in online games (see Figure 10.7). I've found that this is a good feature that does help to balance the gameplay and pair up players who are similarly skilled. There's nothing more frustrating than getting owned over and over again in a game where you are obviously outmatched by just about every other player in the game.

Geek Speak

Owned is slang for being beaten. The one who beats you in a game "owns you," so to speak. A common alternate spelling is p0wn or p0wn3d. Don't ask me why. I'm gen X, not gen Y, so I don't understand most of the L33t5p34k (elite speak) pseudo-language. I was influenced by *Hackers* (the book by Steve Levy) and *Hackers* (the movie by Iain Softley) in my early 20s, so I guess I'm an "01d 1337 h4x0r." (The movie *Hackers* is terribly campy by today's standards, but it gets kudos for introducing the general public to geek culture.) This phenomena is a manifestation of the need to maintain an elitist cultural identity in the midst of outsiders (the *ignorant masses*), according to Wikipedia (http://en.wikipedia.org/wiki/Leet). At the bottom of the Wikipedia page is a link to several L33t5p34k translator programs.

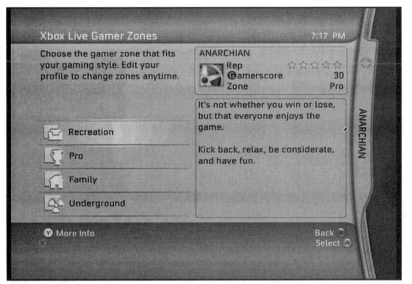

Figure 10.7
Your Gamer Zone determines the types of players you meet.

Gamertag

You can change your Gamertag using the Edit Gamertag screen shown in Figure 10.8. However, the cost is 800 Microsoft Points (approximately $10), which is meant to discourage this practice. Changing your Gamertag does not affect your reputation or Gamerscore, but it requires your friends to add you to their lists again. However, if you delete your Gamer Profile account and

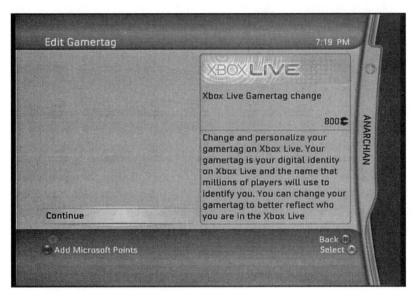

Figure 10.8
You can change your Gamertag—for a price!

create a new one, your gameplay data is lost. There is a partial solution to the problem, though, if you accidentally delete your local Gamer Profile. You can create a new local profile on your 360 based on your Xbox Live account if you are a current subscriber.

Privacy Settings

The Privacy Settings screen (shown in Figure 10.9) allows you to customize the settings for your Xbox Live account to display only the information you want others to see about your account, such as your motto, your played games list, and so on. You can choose exactly what type of information other gamers can see about you and what information you want to keep hidden from them.

Figure 10.9
Editing the privacy settings for your account.

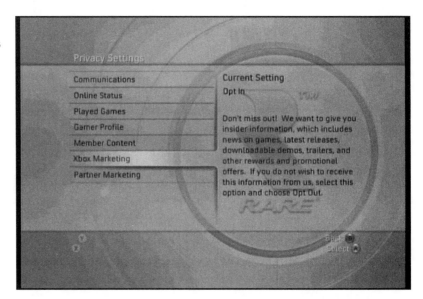

View Rep

Your Gamer Rep is your reputation on Xbox Live. Take it seriously. When you play games with other players, they have the option to review the experience of playing with you, leaving a negative, neutral, or positive feedback, along with some comments. Feedback and comments are optional, but once made, they are permanent and cannot be retracted. The View Rep screen only shows comments after someone has reviewed you (which might be never, because it's not easy to type lengthy messages using the virtual keyboard). If you like to post reviews and think you will use this feature often, you may want to plug in a USB keyboard.

Game Defaults

You can set three different types of game defaults for your gamer profile: general games, action games, and racing games. For general defaults (applicable to game genres other than action or racing), you can set the following options:

- Difficulty
- Primary Color
- Secondary Color
- Controller Sensitivity

Not every game supports these general game defaults, but these options are available should a game developer want to incorporate this feature into a game. These types of settings are the most common ones that you are likely to change in a game, which is why they are offered here as global options for the games that support this feature. (I suspect most games will take a look and heed your customization requests.) Action games have the following options:

- Y-Axis
- Movement
- Auto-Aim
- Auto-Center

Action games include the usual first-person shooters in addition to flight simulations (such as *Battlehawks* and *Crimson Skies*, although these original Xbox games haven't been updated for the Xbox 360 as of yet). The defaults for racing-type games are shown in Figure 10.10 and include the following options:

- Transmission
- Camera Angle
- Brake Control
- Accelerator Control

These options can be helpful if you become tired of configuring every new racing game you get. For instance, do you prefer to use the A button or the right trigger for the throttle? What about the button that engages the main brake versus the parking brake? You can configure at least the main accelerator and brake controls using this interface. Most racing games should use your settings here.

Figure 10.10
Editing the defaults for racing-type games.

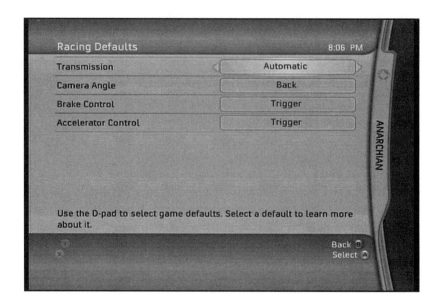

Account Management

The Account Management screen is where you manage your account information, such as your mailing address and passwords. You may need to change your password once in a while (for instance, if your brother or sister "borrows" your gamer profile to play one of your saved games). It's also just plain good practice to change your password every few months, especially if you use an easy password to crack. For instance, I don't recommend using "xbox360" for your password! (Also avoid "god," "sex," and "password," the three most common passwords, according to the movie *Hackers*.)

Disable Auto Sign-In

You can use the Disable Auto Sign-In option to prevent your Xbox 360 from attempting to automatically sign you in to Xbox Live when you power up the system. This may be helpful if more than one person uses the Xbox 360, in which case you would not want someone else playing games online using your Gamertag.

YOUR GAMERSCORE AND REPUTATION

Your Gamerscore represents the number of achievements you have earned while playing games online and offline (in single-player games). Your reputation is determined by other players who rate you after a game round has concluded. This rating system is meant to discourage players from behaving poorly. In practice, most gamers ignore the reputation rating system and simply skip the opportunity to rate others in a game. An exception might be made for someone who swears uncontrollably during a game or who acts like a jerk—in other words, when it is worth the small effort to actually rate another player. Some players take the reputation rating system more seriously than others. I think both Amazon.com and eBay.com have shown how effective a reputation system can be toward filtering out the rotten tomatoes everyone has to deal with in the online world from time to time, so it's natural that Microsoft uses this concept to give players a way to seek other players with good reputations.

Improving Your Gamerscore

You must play a game to see your ranking within that game, because there are no "global" rankings displayed on Xbox Live outside of individual games. By continuing to play the game, you will automatically start climbing the ranks even if you don't win very often, because you will quickly rise above those who play infrequently. To see my world ranking score in *Project Gotham Racing 3*, I need to run the game and select the Leaderboards section of the game for either solo or online career modes. My ranking is shown in Figure 10.11. Because I'm in League 107, I am really far down on the list.

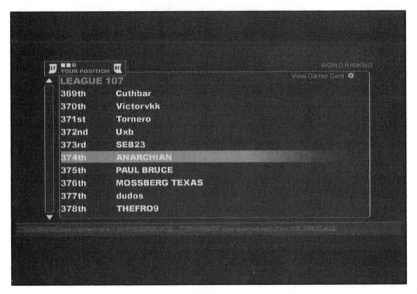

Figure 10.11
Viewing your position in the world ranking.

Because there appears to be 1,000 players in each league, there are more than 100,000 players above me in the ranks! However, this position simply reflects that I have not been playing very long. There are a lot of players in the ranks above me who might have played *PGR3* only a few times before moving on to another game. If you find yourself in a similar position, don't be discouraged. A few weeks of gameplay will help you to quickly climb through the ranks to the upper leagues.

To advance in the ranks, you must reach the first rank in a league to be promoted to the next-highest league (which would be League 106 for me). Let's look at the ranks for League 1, which is the top league in *PGR3*. Figure 10.12 shows the top 10 players of League 1, which represents the best of the best racers in the world—at least in the context of this game.

Figure 10.12
The top 10 ranked players in *PGR3*.

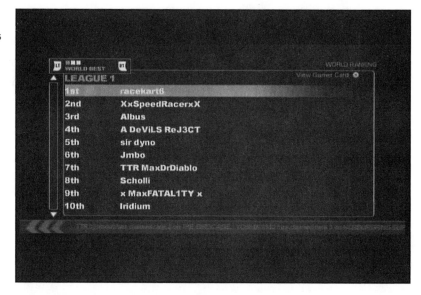

Improving Your Reputation

One of the best ways to rise through the ranks is to play the solo career mode to unlock some of the best cars in *PGR3*. Because I haven't spent much time playing this game yet, I am only racing with the standard cars. This means I'm driving a Dodge Viper SRT10 or a Shelby GT500 against the stereotypical Ferrari F50 GT.

In my opinion, this is a flaw in the game because after you've unlocked the F50 GT, it's the only car you're likely to use in the game, and then it becomes a big competition for rank position—including racing the F50 against newcomers. I might prefer the domestic cars from Ford and Dodge, but these cars are not in the same class as a Ferrari F50 GT. In fact, there is such a huge difference in performance that even the best drivers will be hopelessly left behind driving anything other than another F50 (which is why no one who is concerned with ranking does that). Why are these cars racing together in the same class within *PGR3*? I don't know, but it has led to an imbalance in the game. (Few cars are able to navigate a 90-degree turn at nearly full throttle like the F50.)

The single best way to improve your reputation is by playing fairly with others. There's nothing more frustrating than having someone smash your car into the wall around a tight corner or having a teammate kill you with friendly fire (on purpose or otherwise).

Each game presents its online ranking and high score lists in a unique way, so my example here with *PGR3* will differ from the other games you play online.

PLAYING ONLINE WITH FRIENDS

You can have a lot of fun with your Xbox 360 by playing solo or with your friends using multiple controllers. But the experience will never quite equal playing online against seven other players—or in the case of some games, perhaps 16 or more players in a single game! When you play online, each game can take advantage of the social features built into the Xbox 360 and Xbox Live. For instance, the players in an online game are handled by Xbox Live, and the game coordinates the rendering of the objects on the screen based on information passed to it from the online server. These features are part of the architecture of the Xbox 360 and not something that game developers have to build on their own when programming a game. Figure 10.13 shows the list of players with whom I have recently interacted, with the time that I last saw them online.

Figure 10.13
List of all players whom I have encountered online.

Take a look at Figure 10.14, which shows the results of a race that I just completed in *PGR3*. (This was one of the few times that I've actually won a race, and I suspect it was due to some massive crash early on!) Xbox Live is providing *PGR3* with the list of players and their statistics, such as the order in which the players crossed the finish line (used to determine rank).

Figure 10.14
After you complete a race in *PGR3*, your ranking for the race is displayed.

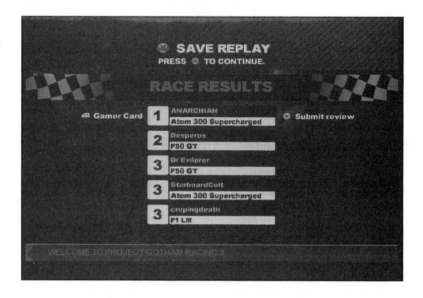

The game displays the results, but it does not do much work as far as processing the actual race, which Xbox Live handles. After you have completed a race, your position in the world ranking is displayed. Figure 10.15 shows that I've moved up a bit. I'm now rising through the ranks in League 104 (up from League 107). All it takes is a few victories to send your Gamertag flying up the ranks toward League 1, because a victory against higher ranked players multiplies the results in your favor.

Figure 10.15
Rising through the world ranking.

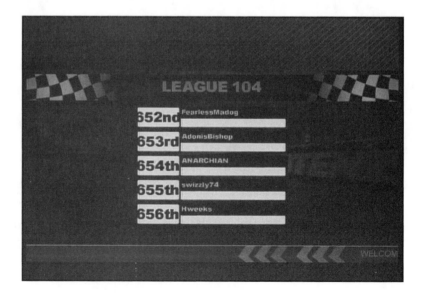

Voice Chat and Instant Messaging

Voice chat is an integral part of Xbox Live and is possible with the voice chat headset (see Figure 10.16). You can select another player from your friends list or request a chat with other gamers using the Messages or Friends options on the main Xbox Live folder on the Dashboard.

Figure 10.16
This headset is used for voice chat on Xbox Live.

To open the online chat feature, open Messages or Friends and select the person with whom you would like to chat. This brings up the Create New screen shown in Figure 10.17. If that player is online and accepts your request, you can chat. This works best with friends on your list, because most players don't want to chat with strangers. Your Xbox 360 keeps track of every player you have played online with in each game, so you don't have to try to remember their Gamertag names.

When you want to send a message, the Select Gamertag screen appears, allowing you to select from the list of players with whom you have played online (see Figure 10.18).

Figure 10.17

Sending messages and chatting with other players.

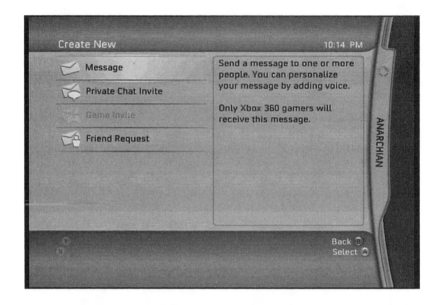

Figure 10.18

Selecting a player as a recipient for your message.

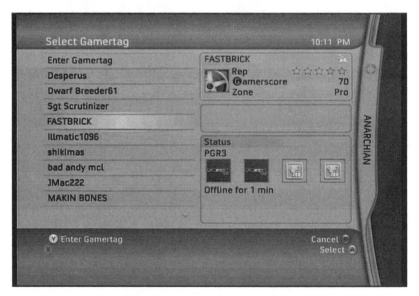

By selecting a Gamertag from the list, you can send that person a text message or a voice message. Figure 10.19 shows the Message screen. If you select the Add Voice option, you can record a voice mail message for that player (up to 15 seconds long). The voice mail is sent to that player's inbox and played back as if it were a text message, but with your actual voice (see Figure 10.19).

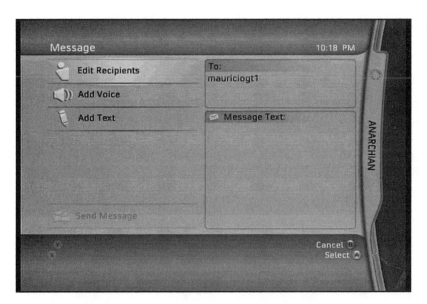

Figure 10.19
Preparing to send a message to another player.

Let's record a voice mail message and then deliver it to another player along with a text message. While that player is doing things in his Xbox 360 Dashboard or playing another game, your message briefly pops up on the screen as a notification. Let's record a message using the Message screen. Choose the Add Voice option to bring up the screen shown in Figure 10.20, and then record your voice mail message. Next, type a short text message to the other player using the

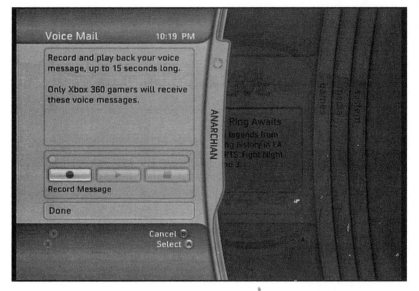

Figure 10.20
Recording a voice mail message for another player.

virtual keyboard, as shown in Figure 10.21. When you've finished typing the text message or recording the voice mail, go ahead and select Send.

Figure 10.21
Typing a text message to send another player.

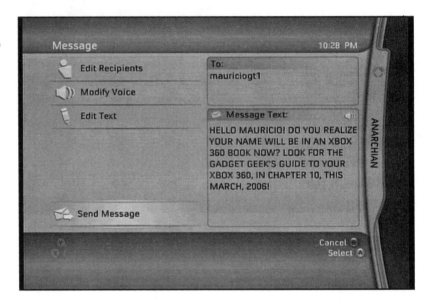

The instant messaging feature of Xbox Live lets you type a message to anyone who is online. That person will immediately see a notification of your message. You can even jump into the Control Panel to send and respond to instant messages sent *to you* by pressing the jewel on your controller to access the chat features. What's more, you can have an active voice chat session with one person (or a group) while playing a game with a completely different group of players.

XBOX LAN PARTIES

The most common form of network gameplay for hardcore Xbox fans is definitely the LAN party. I have hosted or attended about a dozen Xbox LAN parties over the past few years, and they are extraordinarily fun. This type of party was borrowed directly from PC gamers. Some of the earliest LAN parties took place in the mid-1990s after the release of id Software's first *Quake* game in early 1996. By late 1997, the release of *Quake II* resulted in a whole new industry for network gaming, with big companies sponsoring events that eventually led to the formation of the Professional Gamer's League (PGL). This organization sets up national and international game competitions with real prizes and money offered by sponsors for the top players. The Cyberathlete Professional League (http://www.thecpl.com) and Cyber X Gaming (http://www.cyberxgaming.com) are the other two major forces in competition gaming.

The PGL (http://www.thepgl.org) and similar organizations are now involved in a huge tournament industry with massive competitions involving thousands of players hooked up to networks in huge warehouses and convention centers. Huge payouts are offered by sponsors like nVidia, ATI, Abit, Intel, AMD, and other companies involved in the PC industry. The first real "star athlete" of the video game revolution was Johnathan Wendel, aka "Fatal1ty." His insane skill in first-person shooters like *Quake III Arena* and *Counter-Strike* has earned him big bucks through product endorsements for companies like Abit (which sells a "Fatal1ty" motherboard) and Creative Labs (which sells a "Fatal1ty" Sound Blaster).

There are similar competitions for video game consoles like the Xbox and Xbox 360, and huge tournaments take place every year. But for every professional tournament that takes place, there are thousands of private tournaments and LAN parties soaking up megawatts of electricity around the world every weekend.

EPILOGUE

I hope you have enjoyed reading this book and have learned a few things along the way that will empower you to squeeze every ounce of fun from your Xbox 360. I certainly have enjoyed writing this book and learning of the many innovative ideas that went into the design of the Xbox 360. This is truly a video game system that has exceeded every expectation by a wide margin. The next few years will be interesting, to say the least, as Sony and Nintendo attempt to compete with Microsoft's powerhouse next-gen console. If you would like to talk shop, I welcome you to visit the Hardware Hackers and Console forums at http://www.jharbour.com. Or you could try out conferences such as E3, Game Developer's Conference, and Austin Game Conference, all of which I frequently attend. See you there!

the gadget geek's guide to

Your Xbox 360

Index